9 Progress in Clinical Biochemistry and Medicine

Calcitonins – Physiological and Pharmacological Aspects

Mafosfamide – A Derivative of 4-Hydroxycyclophosphamide

Enzymatic DNA Methylation

With Contributions by
M. Azria, J. Engel, S. Grünwald, P. Hilgard,
U. Niemeyer, M. Peukert, G. P. Pfeifer,
J. Pohl, H. Sindermann

With 21 Figures

Springer-Verlag Berlin Heidelberg NewYork
London Paris Tokyo Hong Kong

ISBN-13:978-3-642-74736-6 e-ISBN-13:978-3-642-74734-2
DOI: 10.1007/978-3-642-74734-2

Softcover reprint of the hardcover 1st edition 1989

2151/3020-543210

Editorial Board

Table of Contents

Calcitonins — Physiological and Pharmacological Aspects

Moise Azria

Sandoz Ltd., Pharmaceutical Division, 4002 Basle, Switzerland

Calcitonin is an endogenous regulator of calcium homeostasis, acting principally on bone. The hormone has been isolated from several species of vertebrate, and four types are currently in therapeutic use. Those derived from salmon and eel are more potent than human and porcine calcitonin. Though used primarily to correct various hypercalcemic states, calcitonin is now finding applications in other disorders, and pharmacological studies promise to extend its therapeutic usefulness. The need to administer calcitonin by injection has limited its clinical application, but the recent introduction of an intranasal spray should improve patient acceptability.

1 Introduction

Calcitonin is an endogenous regulator of calcium homeostasis, acting principally on bone. Its hypocalcemic effect was first noted by Copp in 1961 [1,2]; some years later this effect was found to be due to a direct inhibitory action on osteoclast activity.

In 1964, Foster and MacIntyre [3] showed unequivocally that calcitonin was secreted by the thyroid and not, as Copp had thought [4], by the parathyroids. MacIntyre's group further demonstrated that the hormone was produced by the parafollicular cells ("C" cells) of the thyroid [5]. In non-mammalian vertebrates, these cells are located in the ultimobranchial body, which has a common origin with the thyroid gland. Calcitonins have been isolated from various species (ox, chicken, man, salmon, eel, etc.), and five of these have been synthesized. Many synthetic analogues have also been prepared. So far, only four calcitonins are in general medical use — synthetic salmon and synthetic human calcitonins, natural porcine calcitonin and a synthetic derivative of eel calcitonin (aminosuberic^{1,7}-eel calcitonin).

The principal indications for the therapeutic use of calcitonin are disorders involving hypercalcemia, Paget's disease of bone, high-bone-turnover osteoporosis, vitamin-D intoxication, bone metastases and chronic pain associated with bone disease.

Studies of the physiological and pharmacological effects of calcitonin, many of which are not yet fully understood, are leading to new and unexpected therapeutic uses. Calcitonin has, for instance, already found clinical application in the treatment of acute pancreatitis.

2 Structure

The calcitonins are peptide hormones with molecular weights around 3500. The molecule consists of a chain of 32 aminoacid residues with a proline amide group at

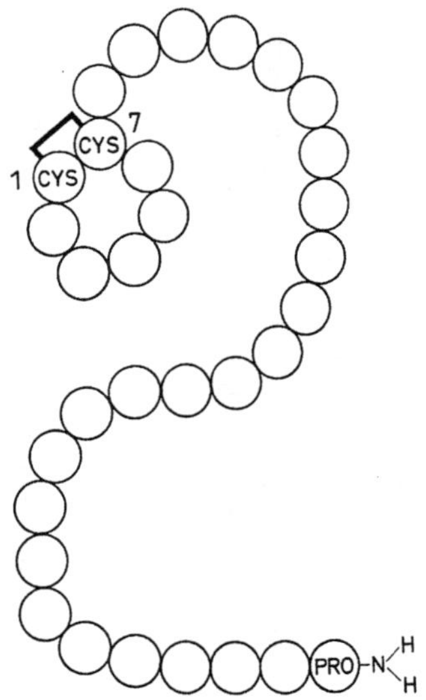

Fig. 1. Basic features of the calcitonin molecule

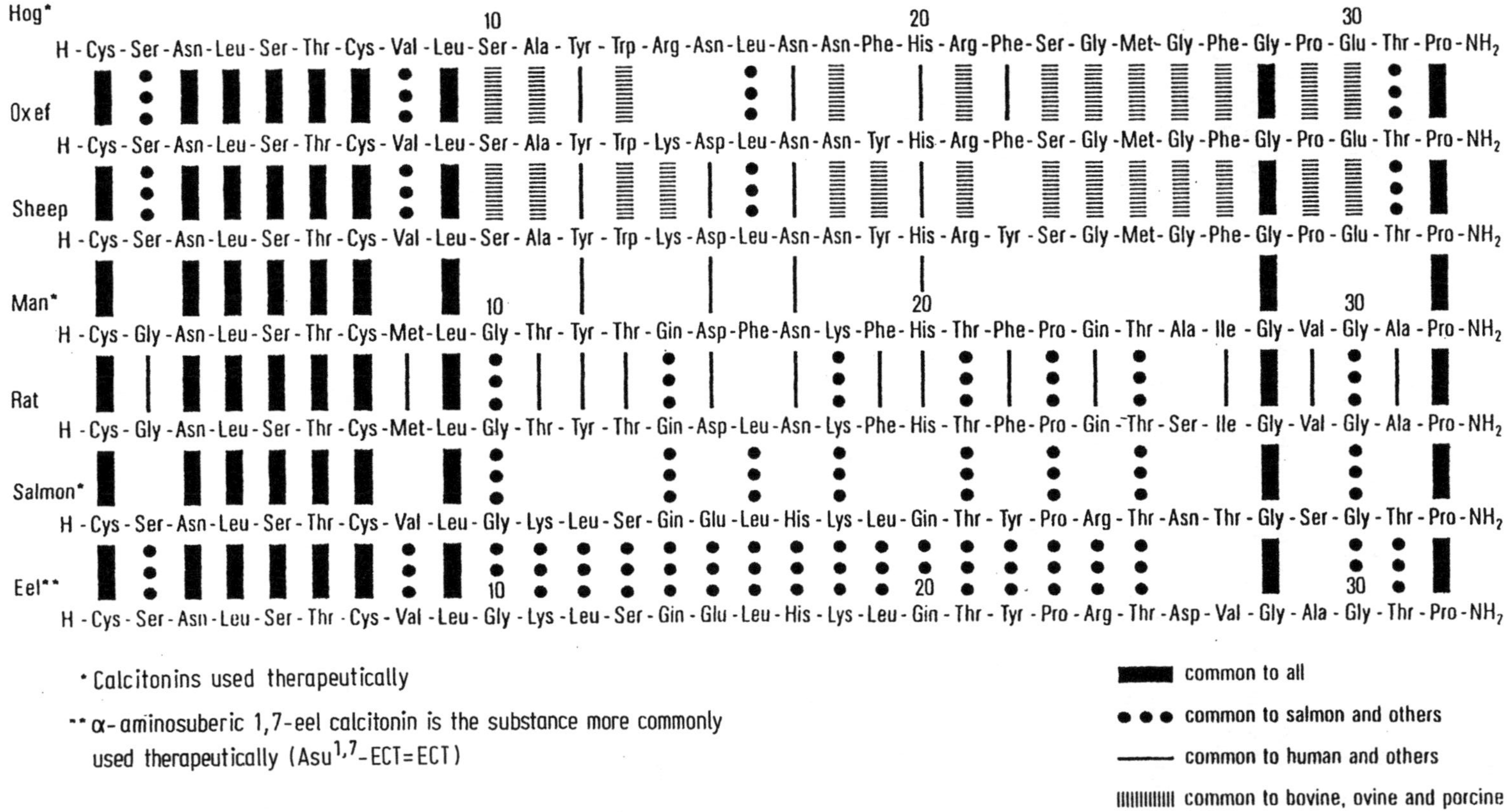

Fig. 2. Comparison of the primary structure of the principal calcitonins

the C terminal. A disulphide bridge between the cysteine residues in positions 1 and 7 forms a ring of 7 aminoacid residues at the N terminal, which carries a free amino group (Fig. 1). The aminoacid composition of the central part of the chain varies from one calcitonin to another (Fig. 2). Similarity of structure is not necessarily reflected in similarity of activity.

3 Assay Methods

There are basically two techniques for calcitonin assay — radioimmunoassay and biological assay.

3.1 Radioimmunoassay

Radioimmunoassay measures immunoreactive calcitonin (iCT), which, however, is not necessarily biologically active. It is a sensitive technique, but its accuracy depends on a number of factors, in particular the specificity of the antiserum used, the limit of detection [6], the presence of analytical artefacts [7–9] and interference by disease processes [10].

Radioimmunoassay is an excellent technique for monitoring blood levels of the hormone and is used to diagnose hypercalcitoninemia associated with medullary carcinoma of the thyroid and, more controversially, postmenopausal hypocalcitoninemic states.

3.2 Biological Assay

Biological assay is based on the measurement of the hypocalcemic response of young rats to a dose of calcitonin. It is less sensitive than radioimmunoassay, but it more accurately reflects true biological activity, or potency. Bioassay cannot be used to distinguish between different types of calcitonin.

The effect of calcitonin on blood levels of calcium is also applied clinically in the acute hypocalcemia test [11], which is used to predict whether calcitonin is likely to be effective in patients with certain bone diseases.

3.3 Other Assay Methods

Other methods used for calcitonin assay, mainly for research purposes, include high-pressure liquid chromatography, radioreceptor assay [12], determination via cyclic AMP levels in plasma, kidneys or urine [13,14], and quantitative cytochemistry [15]

4 Sources of Secretion

The parafollicular, or "C", cells are the primary source of calcitonin secretion [5]. They are concentrated in the thyroid of mammals and the ultimobranchial body of fish, amphibia, reptiles and birds. The C cells are derived from the embryonic neural crest [16,17] from which, during embryonic development, some of them migrate to other sites, where they become secondary sources of iCT. This might explain why calcitonin has been found in the blood and urine of patients after thyroidectomy [18]. However, alternative explanations, such as incomplete extirpation of the thyroid or persistence of receptor-bound calcitonin originating from the thyroid before excision, must also be considered.

Table 1. Sites at which calcitonin has been detected in human tissues by means of an antibody capable of recognizing its carboxyl terminal

Site	Site
Thyroid	Cerebral cortex
Jejunum	Gall bladder
Thymus	Kidney
Bladder	Adipose tissue
Lung	Adrenals
Rectum	Pancreas
Testicle	Spleen
Muscle (skeletal)	Heart
Hypothalamus	Cerebellum
Pituitary	Ovary
Lymph node	Cardiac blood
Liver	Erythrocytes
Stomach	Bone
Oesophagus	Seminal plasma

Table 2. Ectopic sources (tumors) of calcitonin (listed in chronological order of discovery) as detected by either bioassay/RIA or immunofluorescence

Bronchial carcinoid	Renal clear-cell cancer
Intestinal carcinoid	Testicular dysembryoma
Phaeochromocytoma	Myeloma
Stroma ovarii	Oesophageal cancer
Mucosal neuroma	Pleural epithelial cancer
Bronchial carcinoma	Thymic carcinoid
Melanoma	Colonic cancer
Insulinoma	Pancreatic cancer
Hepatoma	Gastric cancer
Mammary cancer	Vipoma
Ganglioneuroma	Pancreatic somatostatinoma
Sympathoblastoma	Paraganglioma
Laryngeal apudoma	

Calcitonin has been detected in, but is not necessarily secreted by, many tissues in the human body [18] (Table 1). The hormone is also secreted by a number of tumors, mostly those in tissues containing C cells [19] (Table 2).

The normal human thyroid contains 1 to 100 μg calcitonin. The amount secreted per day has been reported to be 50–250 μg by one group of workers [20] and 13.8 μg by another [21].

5 Receptors

Binding studies with ^{125}I-calcitonin have demonstrated calcitonin receptor sites in many tissues, including bone [22–24] and kidney cells [22, 23, 25, 26], the central nervous system [27–30] and pituitary gland [28, 31], testicular Leydig cells [32], lymph cells [33, 34] and some types of tumor [35–38] (Table 3).

Table 3. Principal tissues and organs in rat and man containing calcitonin binding sites, or receptors

Bone	Osteoclasts and some marrow cells
Kidney	Proximal and distal convoluted tubules Ascending limb of the loop of Henle Cortical segment of the collecting tubule
CNS and pituitary	*High density of receptors* Hypothalamus Preoptic nucleus and nucleus accumbens Amygdaloid nucleus Zona incerta Interpeduncular nucleus Nucleus griseus periventricularis Reticular formation
	Average or low density Arcuate nucleus and nucleus supramamillaris Substantia nigra Pituitary (pars intermedia and anterior lobe) Spinal cord?
Other sites	Leydig cells Lymph cells (man) Mammary and bronchial tumor cells

6 Levels in Biological Fluids

6.1 Blood Levels

Calcitonin is present in the blood in several forms, most of which are inactive [39–41]. The only biologically active form in man appears to be a monomer identical to synthetic human calcitonin. Reported blood levels in healthy subjects vary widely, and this probably reflects the variable specificity of the antibodies used in radioimmunoassay of calcitonin and the difficulty of the techniques themselves. Quoted levels (expressed as weight per unit volume of immunoreactive calcitonin rather than biological activity) range from 1 to 500 pg/ml plasma [42–46]. 75% of healthy subjects have blood levels below 100 pg/ml [44], the mean level being between 30 and 90 pg/ml [47,48]. Blood levels are an excellent biochemical marker for diagnosing certain diseases and for monitoring the effects of treatment. In view of the apparently wide normal range, blood levels should always be determined with the same assay technique and, if possible, with antiserum of the same specificity.

6.2 Factors Affecting Blood Levels

Although basal levels of calcitonin normally remain fairly constant, they can be affected by various physiological, pharmacological and pathological factors. The principal physiological factors are shown in Table 4.

Table 4. Physiological factors affecting basal calcitonin secretion in man

Factor	Effect on secretion
Circadian rhythm	Rhythmic pattern with periprandial peak at mid-day
Blood level of calcium (and related) ions	Principal regulator of secretion
Food intake	
— Food rich in calcium and protein	↑(declining with age, according to some authors)
— Low-calcium diet	↓
— Alcohol	↑(single shot)
Age	
— Infancy	↑
— Ageing process	↓(no effect, according to some authors)
Sex	Females secrete less than males (no effect, acc. to some authors)
— Pregnancy (esp. in multiparity)	↑
— Lactation	↑
— Raised sex hormone levels	↑
Raised levels of other hormones	
— Hormones of the GI tract	↑
— Gastrin	↑
— Cholecystokinin-pancreozymin	↑
— Glucagon	↑
— Caerulein	↑
— Secretin	↑
— PTH	↑(indirect)
— Thyroxine, iodine	↑(indirect)
— Vitamin D → 1,25 $(OH)_2D_3$	↑
— Prolactin	↑↓(via CNS effect)
Stress	↑

6.2.1 *Ionized Calcium*

The plasma level of ionized calcium, which reflects the rate of bone turnover, is the principal physiological regulator of calcitonin secretion. Rising levels of ionized calcium stimulate calcitonin secretion and release; falling levels inhibit them[49–51].

6.2.2 *Diet*

The body responds to dietary intake of calcium by increasing the quantity of calcitonin available in anticipation of a higher calcium load several hours later. It does this by a mechanism closely related to that regulating the secretion of the hormones of the gastrointestinal tract (gastrin, cholecystokinin-pancreozymin, glucagon, caerulein, secretin), which also stimulate calcitonin secretion[44, 52, 53]. A low-calcium diet is associated with low blood calcitonin levels.

The quantity of calcitonin secreted depends not only on the calcium content of the food eaten but also on its protein content. Calcitonin secretion is also stimulated by consumption of alcohol, which has an adverse effect of its own on bone metabolism.

The stimulatory effect of food intake on the calcitonin response tends to diminish with age, and this diminished response is thought to be the cause of certain bone disorders such as postmenopausal osteoporosis.

Increased calcitonin secretion inhibits the release of calcium from bone into the bloodstream, also giving rise to a slight increase in parathyroid hormone (PTH) secretion, which is not inhibited by the ingested calcium. PTH and calcitonin then act together to regulate calcium homeostasis, PTH reducing the amount of calcium excreted in the urine and calcitonin suppressing the effect of PTH on calcium efflux from bone (Fig. 3).

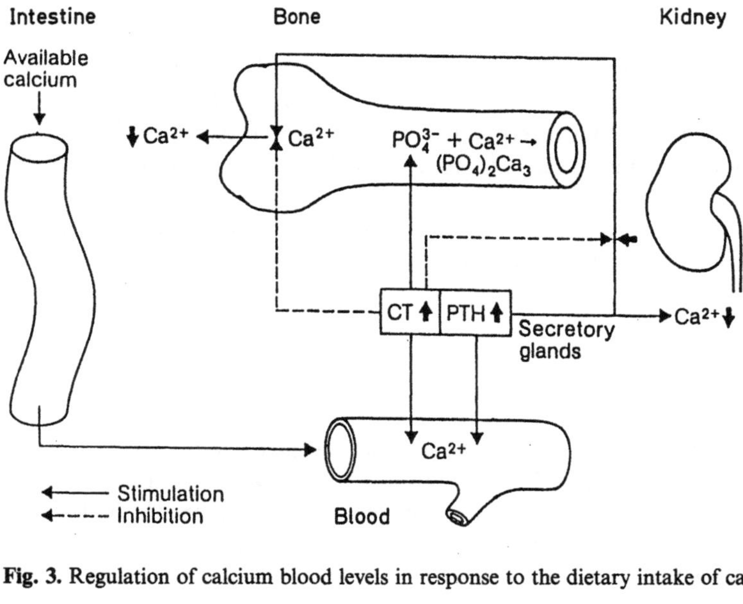

Fig. 3. Regulation of calcium blood levels in response to the dietary intake of calcium

Calcitonin actively moves phosphate into the bone fluid and bone-lining cells and inhibits its return to the bloodstream. Calcitonin directs dietary calcium into bone, where it combines with phosphate and is stored as a temporary reservoir. This reserve is the first source to be called on during a period of food deprivation, thus reducing the need for PTH-induced bone resorption to maintain calcium balance [54)].

6.2.3 *Pregnancy*

Calcitonin secretion increases during pregnancy, probably to protect the maternal skeleton and to meet the demands of the fetus [55, 56)].

6.2.4 *Lactation*

The body's calcium requirement is highest during lactation, when the demands of milk production, of maintaining a steady blood calcium concentration and of safeguarding the integrity of the maternal skeleton have to be met simultaneously. During lactation secretion of three hormones — PTH, calcitonin and $1{,}25(OH)_2D_3$ — increases [57, 58)]. Loss of calcium in the milk leads to a deficiency in the mother's body, which is offset to some extent by enhanced absorption of calcium from the diet. Even so, calcium balance becomes slightly negative, causing a fall in serum calcium, while levels of PTH, calcitonin and $1{,}25(OH)_2D_3$ rise. The interplay of these changes protects the maternal skeleton against excessive calcium depletion while ensuring an adequate calcium supply for the offspring.

The "stress" of lactation appears to be much more intense in rats than in women, presumably because of the rats' larger litter size.

6.2.5 *Age and Sex*

In man basal levels of calcitonin are high at birth and fall with age [47, 59)]; the reverse is true of rats [60)]. Levels are lower in women than in men and fall further after the menopause [61)]. Both estrogens and testosterone stimulate calcitonin secretion [62)], and the postmenopausal fall in calcitonin secretion is attributed to reduced hormone production.

This is the generally accepted view of the effects of age and sex on calcitonin levels. However, both the fall in basal calcitonin levels with age and the sex difference have been questioned [62, 63)].

6.2.6 *Other Physiological Factors*

Physical exertion is followed by an increase in blood levels of calcitonin [64)], as is a rise in the blood level of the vitamin D metabolite, $1{,}25(OH)_2D_3$. When the concentration of $1{,}25(OH)_2D_3$ rises, as during growth, pregnancy or lactation [56)], the level of calcitonin rises to protect the skeleton and concentrate the action of vitamin D and its metabolites on the gut.

6.2.7 *Pharmacological Factors*

Many pharmacologically active substances can affect blood levels of endogenous calcitonin (Table 5). Ethanol [48], sulpiride [65], domperidone [65], pentagastrin [66], isoproterenol [67] and oral contraceptives [68] stimulate calcitonin secretion, whereas cimetidine [65], nicotine [64] and somatostatin [67] inhibit it.

Estrogens are generally, though not universally [68], thought to increase blood levels of calcitonin [61,69], and this suggests how estrogen deficiency could be a factor in the pathogenesis of esteoporosis in postmenopausal and ovariectomized women low calcitonin levels lead to reduced matrix formation, poor mineralization and removal of the brake on resorption, resulting in loss of bone. This is why estrogens are prescribed (together with a progestagen to eliminate the risk of endometrial carcinoma) to prevent osteoporosis in women without functioning ovaries [70–73].

Table 5. Some pharmacological factors affecting basal calcitonin secretion in man

Factor	Factor
Substances stimulating secretion	*Substances inhibiting secretion*
Ethanol	Cimetidine
Sulpiride	Nicotine
Domperidone	Somatostatin
Pentagastrin	Calcitonin (exogenous)(?)
Isoproterenol	
Estrogens(?)	*Substances having no effect*
Testosterone	Dopamine
Oral contraceptives	Levodopa
Calcium (i.v. infusion)	Metoclopramide
Catecholamines	Chlorpromazine
Vitamin D	Bromocriptine
Opioids (in heroin addicts)	TRH

6.2.8 *Pathological Conditions Associated with Depressed Blood Levels*

Some forms of *osteoporosis*, including postovariectomy and postmenopausal osteoporosis, may be associated with calcitonin deficiency [74,75]. This may also be true for osteoporosis secondary to an endocrine disorder [hypogonadism [76] and hypothy-

Table 6. Pathological states associated with depressed calcitonin secretion in man

Osteoporosis
— Postmenopausal/postovariectomy
— Senile
— Secondary to endocrine disorders
Non-goitrous congenital cretinism
Acromegaly(?)

roidism [77)]. Secretion is also depressed in non-goitrous congenital cretinism [78)] and, probably, in acromegaly [76)].

Thyroidectomy is also followed by low blood levels of calcitonin.

6.2.9 Pathological Conditions Associated with Raised Blood Levels

A number of *tumors* secrete calcitonin (Table 2), most notably medullary thyroid carcinoma [79)]. Some extrathyroid tumors also secrete calcitonin, and most of these occur in C-cell-containing tissue derived from the neural crest [80)] or in tissue belonging to the APUD (Amine Precursor Uptake Decarboxylation) system [81,82)]. Release of calcium from bone metastases may also lead to raised blood levels of calcitonin [83)].

Hypercalcemia and primary *hyperparathyroidism* are also associated with raised calcitonin levels, since calcium is the principal stimulant of calcitonin secretion.

The consensus is that basal levels of immunoreactive calcitonin are raised in both acute and chronic *renal failure* [84–90)], although some workers have reported depleted levels in early-stage disease and in juvenile renal failure [91)]. In chronic renal failure calcitonin is secreted in excess in response to raised serum levels of calcium or gastrin [92)] and the metabolic clearance of calcitonin is diminished [87,93,94)].

Both raised and normal levels of calcitonin have been reported in patients with *pancreatitis* [84,95–103)]. Many authors claim that glycogen stimulates calcitonin secretion [84,96,99–101)]; others think that a substance with calcitonin-stimulating activity may be produced in the pancreas itself or in the adjacent gut [81,103)]. On the other hand, it is possible that the inflamed pancreas secretes a substance that interferes with radioimmunoassay.

Heroin addiction is associated with raised levels of calcitonin [104)], even in the absence of impaired kidney and liver function and abnormal serum calcium and phosphate levels. This suggests that calcitonin secretion is stimulated directly by heroin, and possibly by other opioids.

Table 7. Pathological states associated with calcitonin overproduction and raised blood levels

Conditions associated with calcitonin overproduction	*Other conditions with raised calcitonin levels*
Calcitonin-secreting tumors	— Graves' disease
— Medullary carcinoma of the thyroid	— Atrophic gastritis
— Other tumors	— Acute gastritis
	— Pernicious anemia
Hypercalcemia	— Peptic ulcer
Neonatal hypocalcemia	— Gastrointestinal bleeding
Primary hyperparathyroidism (controversial)	— Stress
Pseudohypoparathyroidism	— Thyroid surgery
Renal disorders	— Hepatic surgery
— Renal failure (acute and chronic)	— Toxic shock
— Renal dialysis (esp. high-molecular-weight forms of calcitonin)	— Myocardial infarction
	— Difficult labor
	— Trauma
Pancreatitis	— Lithium intoxication
Heroin addiction	

6.3 Diurnal Variation in Blood Levels

Diurnal variation in calcitonin secretion has been reported, with a peak between 12 h and 13 h [48], but other evidence suggests that there is no diurnal variation in serum concentrations of either calcitonin or $1{,}25(OH)_2D_3$ [105, 106].

6.4 Levels in Cerebrospinal Fluid

In one study calcitonin was found in the cerebrospinal fluid of 75% of a group of 63 men. The mean level was 11.1 ± 1.3 (SD) pg/ml, with a range of <2–55 pg/ml. The plasma: CSF ratio was 2.3:1 [107].

6.5 Levels in Milk

Levels of immunoreactive calcitonin in human milk have been found to be 10 to 40 times higher than serum levels [108], suggesting local production of iCT within the mammary gland or a specific transport and concentration mechanism. Milk levels averaged 2.46 ng/ml 1 week post partum and fell rapidly within a few days (mean level at 3 months: 0.58 ng/ml).

7 Physiological Functions

The principal physiological role of endogenous calcitonin is the regulation of calcium metabolism, chiefly by helping the body to deal with episodes of "calcium stress", e.g. preventing calcium excess.

Calcium, as one of the major constituents of the body's internal milieu, plays a vital part in the maintenance of both structural (skeletal and muscular) and other (endocrine, nervous and circulatory) systems. It is involved, for example, in the control of cellular permeability, neuromuscular excitability, muscular contraction, the activation of certain enzymes (lipase, succinyl dehydrogenase, trypsinogen, ATP-ase), in endocrine secretion, in cardiac function and in blood coagulation [64]. Calcium also directly or indirectly controls the movement of other mineral ions, such as phosphate and magnesium, as part of the process of maintaining ionic equilibrium.

As a hormone of mineral regulation, calcitonin exerts effects at many levels.

7.1 Effects on Bone

At times of calcium stress — e.g. during growth, pregnancy, lactation and after eating — calcitonin protects the skeleton by inhibiting osteoclast activity and proliferation, thereby reducing bone resorption and remodelling [109].

Bone undergoes continual renewal through the activity of two types of cells — osteoclasts and osteoblasts — which together form a remodelling unit [110]. Osteo-

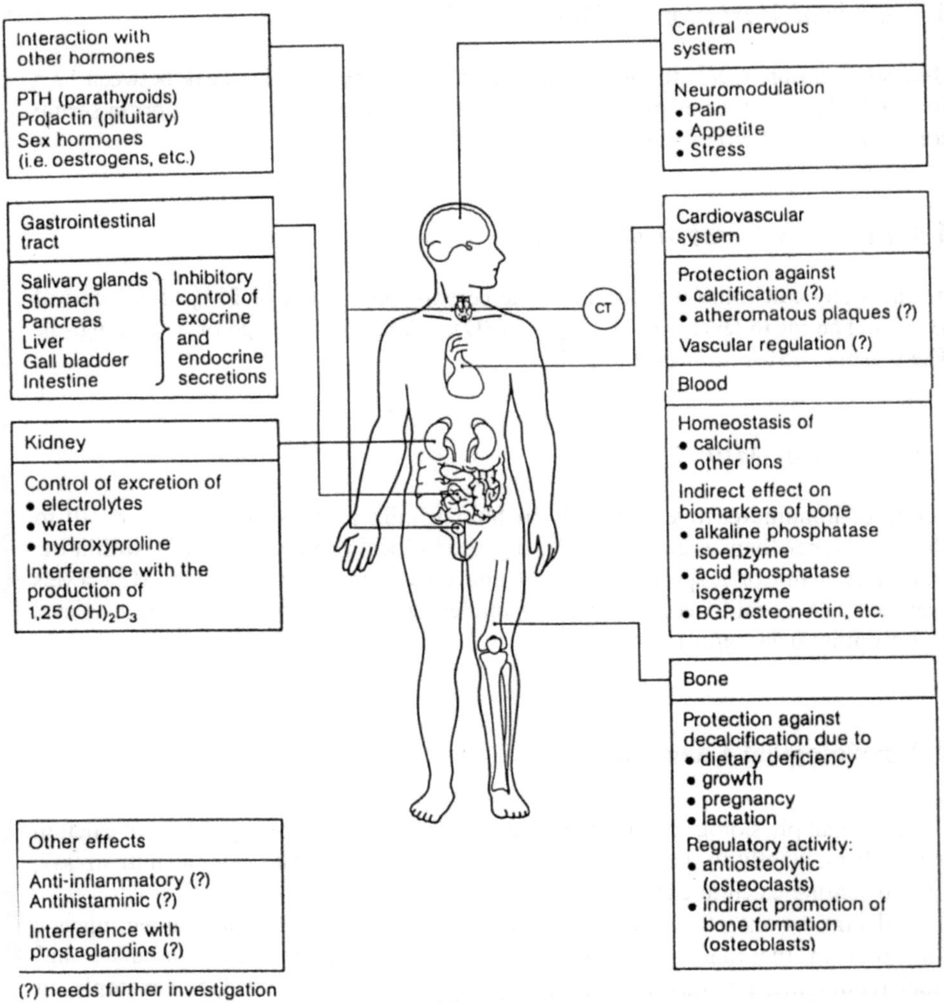

Fig. 4. Diagram showing the main physiological actions of endogenous calcitonin

clasts, which probably derive from the same precursor cells as monocytes [111, 112], are highly mobile cells [113] which move along the bone surface, excavating pits in the bone. The mechanism involved is poorly understood but probably entails resorption of the matrix by a specific process of extracellular digestion by proteolytic enzymes or a collagenase along the brush border in an acid environment. Osteoblasts, which probably derive from precursor cells common to fibroblasts [114–117], secrete the protein matrix of bone and become trapped within it as inorganic salts are deposited. They become transformed into osteocytes, the cells that make up 90% of the cells in bone.

Several agents are involved in bone formation and resorption:

Calcitonin inhibits osteoclast activity and proliferation. This slows bone resorption, thus indirectly lowering plasma calcium levels [118, 119]. It may also exert an effect on osteoblasts, either indirectly via the osteoclasts or directly by stimulating bone formation [109, 120, 121].

Parathyroid hormone (PTH) stimulates osteoclast activity, thereby exerting a potent, if indirect, hypercalcemic effect. It can also stimulate the proliferation, or conversely inhibit the activity, of osteoblasts, depending on its concentration [122].

Vitamin D metabolites, notably $1,25(OH)_2D_3$, play a role by increasing intestinal absorption of calcium.

Estrogens slow bone resorption, possibly indirectly by stimulating calcitonin. However, there is some evidence for the existence of estrogen receptors on bone-related cells (osteoblasts) and this suggests that estrogens may exert a direct effect on bone [123–125].

Corticosteroids reduce osteoblast activity and thus bone formation.

Thyroid hormone appears to stimulate both osteoclasts and osteoblasts.

Calcitonin is thought to control osteoclast activity by regulating, together with PTH, the movement of calcium between the extracellular, intracellular and mitochondrial compartments. Inside the cell calcium may remain in the free state in the cytosol or be reversibly deposited in the mitochondria [126–128]. Calcitonin is claimed to promote this deposition, an effect that is enhanced by phosphates, which increase the ability of mitochondria to accumulate calcium [129, 130]. PTH, on the other hand, indirectly promotes calcium efflux from the mitochondria into the cytosol and thence in the direction of the extracellular fluid [131–133]. This redistribution of calcium during osteolysis is accompanied by increased release of calcium from bone and increased reabsorption by the renal tubules. The net gain or loss in total cell calcium depends on the concentration ratio of intracellular to extracellular calcium [124, 131–134].

Some authors suggest that the primary effect of calcitonin on bone mineralization is to regulate the uptake of phosphate by bone cells, the combination of phosphate and calcium leading to precipitation of hydroxyapatite. According to this hypothesis the effect of calcitonin on blood calcium is simply a consequence of mineral nucleation after phosphate uptake [128, 135] and is either distinct from its effect on bone resorption or else linked via the inhibitory action of phosphate on resorption [136].

7.2 Effects on the Kidney

Calcitonin appears to play a minor, though direct, part in normal renal function, being involved to some extent in electrolyte and water excretion, and consequently in calcium homeostasis. The cortex of the human kidney contains a calcitonin-linked adenylate cyclase, guanylate cyclase, and $Ca^{2+}Mg^{2+}$-dependent ATPase. At low doses calcitonin might also enhance 1-α-hydroxylase activity in the proximal straight tubule, which increases the production of $1,25(OH)_2D_3$ from its substrate [137, 138].

7.3 Effects on Calcium Homeostasis

The function of all movements of calcium is to maintain a normal blood level during episodes of calcium stress. Removal of C cells by thyroidectomy impairs the homeostatic response to hypercalcemia induced by calcium, PTH or vitamin D, whereas if

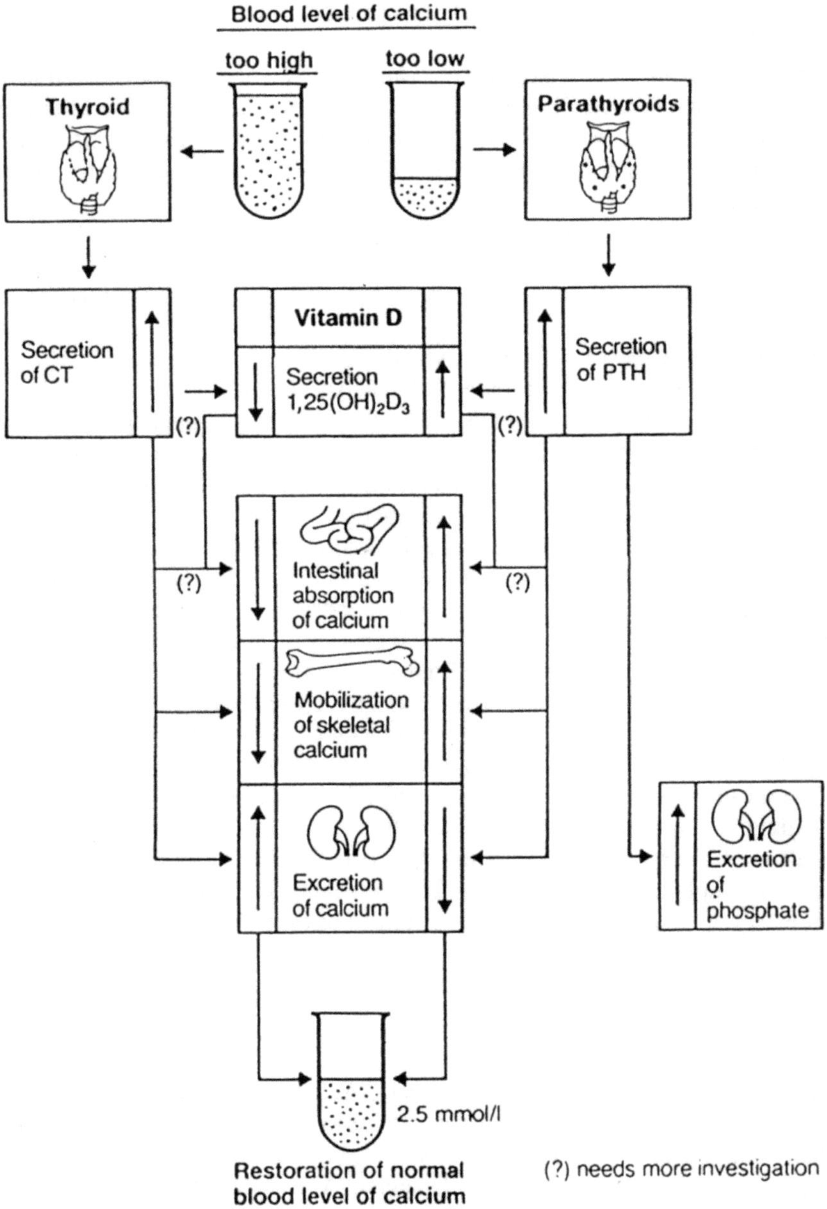

Fig. 5. The principal mechanisms regulating calcium homeostasis

the thyroid is intact, calcitonin secretion automatically increases in response to such stress, so as to maintain calcium balance [127, 128]. The principal mechanisms of calcium homeostasis are shown in Fig. 5.

Calcium is the fifth most abundant element in the body, most of it being stored in bone. The normal plasma level is approximately 2.5 mmol/l; 40% of this is bound to proteins (albumin), 10% is diffusible but complexed with anions such as citrate and phosphate, and the remainder is in diffusible ionized form. This last is the physiologically most active form and the one that should be assayed before the diagnostic significance of a given plasma calcium value can be safely interpreted.

Renal reabsorption of calcium is stimulated by PTH and vitamin D via effects on the distal and proximal tubules, respectively.

Serum calcium levels much above or below the range 2.25–2.65 mmol/l indicate an abnormality of calcium metabolism.

Hypocalcemia may result from hypoparathyroidism or a severe deficiency of vitamin D [84], but it is never due to a lack of dietary calcium, although there is no agreement on this point. Repeated transfusions of citrated blood may also result in temporary hypocalcemia [139].

Hypercalcemia may result from hyperparathyroidism, raised blood levels of $1,25(OH)_2D_3$, hypophosphatemia or malignancy. Serum calcium levels above 3.25 mmol/l may cause renal failure due to massive deposits of calcium salts in the kidney, mental disturbances (such as confusion, lethargy), cardiac arrhythmias, calcareous aggregates in soft tissues and diminished response to analgesics.

7.4 *Effects on the Gastrointestinal Tract*

Secretions of the gastrointestinal tract depend on numerous factors, among them the local calcium concentration. Therefore they depend on the activity of calcitonin. That calcitonin affects the functioning of both endocrine and exocrine glands in the gastrointestinal tract is supported by the following evidence:

1) C cells and the secretory cells of the gastrointestinal tract have a common origin [82, 140, 141].
2) The disturbance of calcium metabolism seen in primary hyperparathyroidism is often accompanied by the triad of peptic ulcer, pancreatitis and cholelithiasis [142, 143].
3) Induced hypercalcemia increases the acidity of gastric juice in normal subjects [142, 143].
4) Several gastrointestinal hormones are potent calcitonin secretagogues [44].
5) The presence of calcium ions at normal local concentration is essential for insulin secretion [144].

7.5 *Effects on the Central Nervous System*

Calcitonin (or an analogue) and calcitonin receptors are present in the central nervous system in zones involved in the control of pain perception, appetite, prolactin secretion and lactation [29, 31, 145–148]. The hormone and its binding sites are found in the hypothalamus, possibly suggesting that calcitonin has a neuromodulator function [149].

7.6 Other Physiological Roles

Other roles claimed for calcitonin include anti-inflammatory activity, prevention of calcium deposition in the cardiovascular system and of the formation of atheromatous plaques, an antihistaminic effect, interaction with prostaglandins and an anti-stress effect [64]. These effects are minimal at physiological levels but may be more important at pharmacological levels. The common denominator in the various physiological actions of calcitonin could well be the modulation of calcium flux across the membranes of different types of cells and thus of the intracellular-extracellular distribution of calcium in various systems [150, 151].

8 Biochemical Markers

Biochemical factors associated with bone turnover and calcitonin activity may be used as markers in physiological, pathological and therapeutic situations. The principal markers are blood levels of cyclic AMP [152], alkaline phosphatase, acid phosphatase, hydroxyproline [153, 154], and bone GLA protein (BGP or osteocalcin) [155–161].

9 Pharmacology

The effects of pharmacological doses of calcitonin are wide-ranging and suggest that the hormone has therapeutic potential beyond its primary role in calcium regulation.

9.1 Effects on Bone

The pharmacological action of calcitonin on bone is primarily an inhibition of bone resorption, which it achieves by reducing the activity and number of osteoclasts [162]. This anti-osteolytic effect is thought to retard bone demineralization and breakdown of the matrix. By depressing osteoclast activity, calcitonin may also indirectly depress osteoblast activity. This could explain its apparent success in Paget's disease in normalizing bone turnover.

Calcitonin might also promote bone formation through stimulation of osteoblasts, chondrogenesis and matrix mineralization [163]. It is also possible that calcitonin indirectly prolongs the formation phase of the bone remodelling unit's cycle. However, these two hypotheses are still controversial and need further investigation.

9.2 Effects on the Kidney

High doses of calcitonin have a direct and rapid effect on the kidney, increasing urinary levels of calcium, phosphate, magnesium, sodium, potassium and chloride [164–167]. This effect is seen both in patients with bone diseases and in healthy subjects [167] and is independent of parathyroid activity [168]. The mechanism probably involves stimulation of specific adenylate-cyclase-linked receptors [169], which trigger an increase in cAMP production [170–173]. Urinary levels of hydroxyproline, which are raised in conditions of intensive bone remodelling activity, are lowered by long-term treatment with calcitonin [174].

9.3 Effects on the Gastrointestinal Tract

The principal effect of calcitonin on gastrointestinal function is inhibition of gastric and pancreatic secretory activity. This suggests that it might have potential in the prevention, and even the treatment, of peptic ulcer. At very high doses calcitonin increases the secretion of water, sodium, potassium and chloride ions into the intestinal lumen [175–177]. This effect is negligible at normal therapeutic doses, but it might explain the diarrhea sometimes associated with high doses [178]. It might also account for the diarrhea which affects patients with medullary carcinoma of the thyroid but which disappears after tumour resection [179]. At doses used in bone diseases calcitonin does not appear to affect intestinal calcium absorption. At slightly higher doses it relieves pain in acute pancreatitis and restores enzyme production to normal levels [180–184].

9.4 Effects on Blood Chemistry

The principal effects of calcitonin on blood chemistry are hypocalcemia and hypophosphatemia. The magnitude of the *hypocalcemic effect* (Fig. 6) depends on the dose

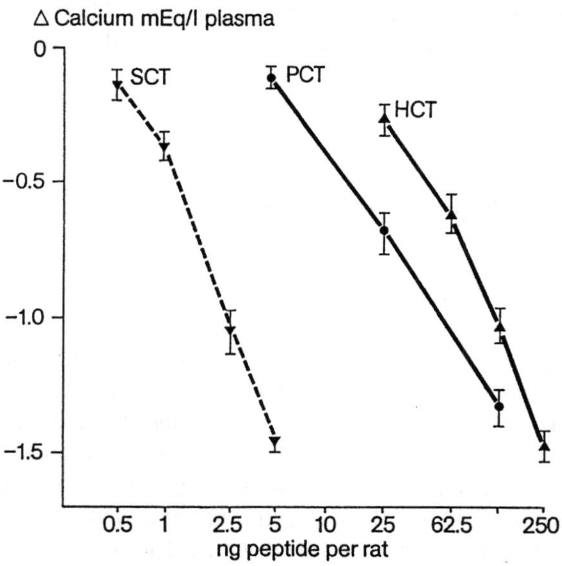

Fig. 6. Dose/hypocalcaemic response curves for human (HCT), pig (PCT) and salmon (SCT) calcitonins, as determined 50 minutes after a single intravenous injection in young rats. Each point represents the mean of 10 rats ± s.e. mean

used [185,186], the concentrations of other ions present (especially phosphate), and interactions with other hormones, including PTH. The effect also depends on the level of bone remodelling activity [187]. In healthy adults the hypocalcemic response is slight, blood calcium levels falling by only 3 to 5 mg/l at 1 to 4 hours after administration. When bone remodelling activity is high, as in children and patients with certain bone diseases, blood calcium may fall by as much as 15 mg/l.

The *hypophosphatemic effect*, which results from a combination of a lower rate of bone resorption and a hyperphosphaturic effect, increases with dose [185,188].

9.5 Effects on the Central Nervous System

The principal effect of calcitonin on the central nervous system is analgesia [189]. It appears to relieve both bone pain [due to tumor metastases [190,191], Paget's disease [192], and osteoporosis [193]], and pain not associated with disorders of bone [194]. How this effect is mediated is uncertain. Suggested mechanisms include an effect on calcium flux in the neuronal membrane [189], an action at specific central receptors [195], an increase in β-endorphin levels [193,196], inhibition of prostaglandin synthesis [197], or simply an indirect result of general improvement in painful bone lesions, resulting in reduced pain perception at central level.

The route of administration is the most important factor in determining the time of onset of calcitonin-induced analgesia, being shortest after subarachnoid administration. The more peripheral the site of injection, the longer the delay and the larger the dose required to produce an analgesic effect. Human calcitonin has a short-acting analgesic effect, whereas the effect of salmon calcitonin lasts 6–8 hours [195].

Other putative effects, such as appetite suppression, are unconfirmed. However, all the evidence suggests that calcitonin has a neuromodulatory function in the central nervous system [198].

9.6 Interactions with Other Endogenous Hormones

9.6.1 Hormones Regulating Calcium Metabolism

Calcitonin stimulates 1-α-hydroxylase activity, therby increasing production of $1{,}25(OH)_2D_3$, the active metabolite of vitamin D [199,200]. Calcitonin administration is followed by an immediate rise in blood levels of *parathyroid hormone* [201], its functional antagonist in the regulation of blood calcium. Their interaction is indirect, being mediated by the blood level of calcium. The activation of adenylate cyclase in bone and kidney cells by PTH is not blocked by calcitonin, and the parathyroid gland continues to function normally during long-term treatment with calcitonin [202,203]. Hyperparathyroidism is not a typical feature of medullary carcinoma of the thyroid [204]. Moreover, the two hormones act at different receptors in bone and kidney [162,172,205].

9.6.2 *Hormones of the Gastrointestinal Tract*

Calcitonin appears to stimulate production of *vasoactive intestinal polypeptide* (VIP) (which is thought to cause the flushing often seen as a side-effect of calcitonin treatment [178)]) and *somatostatin* [206)]. Calcitonin inhibits the secretion of *insulin* [207–209)] and *gastrin* [207, 210, 211)].

9.6.3 *Anterior Pituitary Hormones*

In man secretion of *thyrotrophin* [212, 213)], *luteinizing hormone* [212, 213)] and *growth hormone* [207, 214)] is slightly inhibited by calcitonin. In healthy subjects and patients with acute episodes of osteoporosis, calcitonin significantly raises levels of *corticotrophin* (ACTH) [196)]. Calcitonin inhibits *prolactin* secretion [215, 216)] via a direct action on the dopaminergic system [217)].

9.6.4 *Beta-Endorphin*

Central and peripheral levels of β-endorphin are reported to be raised after calcitonin administration, producing an analgesic effect [193, 196)].

9.7 *Anti-Inflammatory Properties*

Calcitonin shows anti-inflammatory properties in experimental models of acute and chronic inflammation in rats and mice. This effect is dose-dependent. In the carrageenan rat paw edema test, salmon calcitonin is three times as potent as human and porcine calcitonin. The anti-inflammatory mechanism probably involves a reduction in vascular permeability [218)] and stimulation of lysine decarboxylase activity, as with many non-steroidal anti-inflammatory agents. Inhibition of prostaglandins, and thromboxane biosynthesis has been observed in animals but not, so far, in man [219)].

9.8 *Cardiovascular Effects*

Calcitonin has no clear-cut major effects on the cardiovascular system. The most notable effect is the flushing often seen as a side-effect of calcitonin treatment, probably mediated by release of VIP. Long-term administration of porcine calcitonin reduced the extent of experimental arteriosclerosis in rabbits [220, 221)], presumably by inhibiting the formation of calcareous deposits in the arteries (atheromatous plaques) by preventing the entry of calcium ions into smooth muscle.

9.9 *Anti-Stress Effect*

Stress commonly gives rise to increased blood levels of calcitonin, and a protective effect against stress has been attributed to the hormone [222)]. The fact that the hypothalamus, the control centre of the autonomic nervous system, contains a high density of calcitonin receptors supports this suggestion.

The anti-ulcerogenic effect of calcitonin (in animals) [223] can be regarded as an anti-stress effect.

10 Pharmacokinetics, Bioavailability and Metabolism

The principal calcitonins in therapeutic use are synthetic salmon calcitonin (SCT), synthetic human calcitonin (HCT), natural porcine calcitonin (PCT) and the 1,7 amino-suberic derivative of eel calcitonin (ECT). Only parenteral routes of administration (intravenous, intramuscular, subcutaneous and, as of very recently, intranasal) are feasible at present.

SCT and ECT have 20–40 times greater biological activity than HCT and PCT (Table 8). The metabolic clearance rate of SCT in only 2–3 times lower than that of HCT and PCT, and its greater potency is likely to be due to a higher level of intrinsic activity at its receptors.

SCT, the calcitonin most extensively studied [224], is efficiently absorbed after intramuscular and subcutaneous injection, and peak plasma levels are attained in less than an hour (Fig. 7). The half-life of elimination from plasma is between 70 and 90 minutes, the apparent volume of distribution after subcutaneous injection about 20 litres, the metabolic clearance rate slightly over 200 ml/min and plasma protein binding 30–40%. The bioavailability of SCT after both subcutaneous and intramuscular injection is about 70% [225], and this is not improved by addition of gelatin to the vehicle. Only about 2% of the administered dose is excreted in urine as unchanged hormone.

The calcitonins are degraded mainly at extravascular sites [226–228], the liver in the case of PCT and the kidney in the case of HCT, SCT and ECT. SCT and ECT are more resistant to metabolic degradation than PCT and HCT and are eliminated more slowly.

No correlation has yet been found between plasma levels and clinical efficacy. Although the peak plasma level is reached within an hour of administration, the

Table 8. Hypocalcaemic activity of the calcitonins

Species	Activity (IU/mg)	Activity (mg/unit)
Salmon I Salmon II Chicken $Asu^{1,7}$-eel	4000–6000	0.00017–0.00025
Salmon III Eel	2000–4000	0.00025–0.0005
Rat	≃400	≃0.0025
Ox Sheep Pig Man	100–200	0.005–0.01

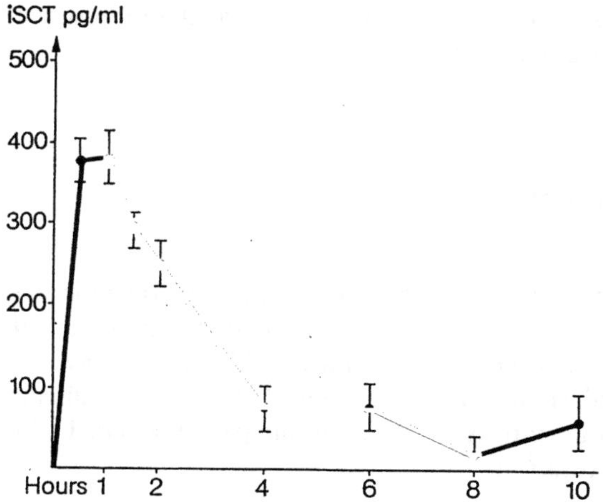

Fig. 7. Mean (±SEM) SCT plasma level curve for 16 healthy volunteers given a dose of 19.9 μg s.c.

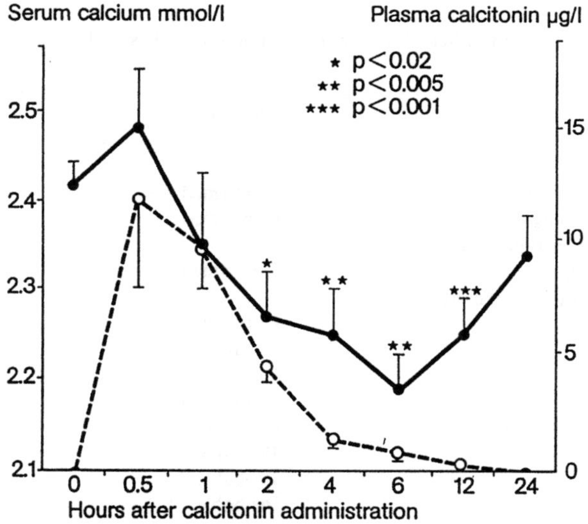

Fig. 8. Serum calcium (—) and plasma calcitonin (---) levels (mean ±SEM) in 7 pagetic patients after a single dose of synthetic HCT 100 IU s.c.

maximum hypocalcemic effect occurs later and sometimes persists after the blood level has fallen below the threshold of detection (depending on the assay method used) (Fig. 8).

Endogenous calcitonin has an elimination half-life of about 10 minutes in man, with the kidney as the principal site of metabolic clearance.

10.1 Pharmacokinetic Data as a Guide to Dosage

Pharmacokinetic data obtained by radioimmunoassay tend to be lower than those obtained by radiolabelling, and monitoring of plasma levels is not a reliable way of determining dosage requirements, especially in severe hypercalcemia. Dose adjustment should preferably be based on clinical measurements.

11 Calcitonin in Therapeutic Use

The introduction of synthetic forms of calcitonin has facilitated its therapeutic use. The many disorders in which calcitonin is therapeutically useful are listed in Table 9. Established indications include hypercalcemic states with or without bone involvement, such as bony metastases and vitamin D intoxication [229, 230], Paget's disease of bone [231], high-bone-turnover osteoporosis [232–235], chronic pain associated with bone disease [236] and acute pancreatitis [237].

Table 9. Therapeutic indications for calcitonin either alone or in conjunction with specific causal treatment

Indication	Efficacy
I Hypercalcaemia	
— Hypercalcaemic crises, emergency treatment	Confirmed
— Due to primary hyperparathyroidism	Confirmed
— Due to hyperparathyroidism of neoplastic origin with or without bone metastases	Confirmed
— Due to multiple myeloma or other malignant blood cell disorders	Confirmed
— Due to thyrotoxicosis	Confirmed
— Due to immobilization	Confirmed
— Due to vitamin-D intoxication	Confirmed
— Due to drug effects (thiazide diuretics)	Confirmed
— Due to milk alkali syndrome	Probable
— Due to acute renal failure	Probable
— Due to Addison's disease	Probable
II Bone disease with decalcifying lesions	
A Generalized osteoporosis (prevention/treatment)	
— Senile	Confirmed
— Endocrine	
• Postmenopausal	Confirmed
• Hypogonadic or following castration	Confirmed
• Cushing's disease	Probable
• Hyperthyroidism	Probable
• Acromegaly	Possible
— Congenital	
• Osteogenesis imperfecta	Confirmed
• Juvenile	Probable
— Associated with Gl and digestive processes	
• Postgastrectomy	Possible
• Cirrhosis of the liver	Possible
• Steatorrhoea	Possible

Table 9. (continued)

Indication	Efficacy
— Associated with neurodystrophy	
• Immobilization	Confirmed
• Neuropathy	Confirmed
— Following irradiation	Probable
— Associated with diabetes	
• Diabetes mellitus	Possible
• Haemochromatosis	Possible
— Iatrogenic	
• Heparin	Confirmed
• Cortisone	Confirmed
• Thyroxine	Confirmed
B Localized post-traumatic osteoporosis	
Sudeck's atrophy	Confirmed
— Periodontosis	Probable
C Mixed forms (with bone destruction, decalcification and remodelling)	
— Paget's disease	Confirmed
— Metastases of prostatic carcinoma	Probable
— Metastases of other carcinomas	Probable
— Renal osteodystrophy	Confirmed
D Pain associated with bone disease	
— Chronic bone pain in osteoporosis, bone lesions of neoplastic origin, and other bone lesions	Confirmed
III Diseases of the stomach and pancreas	
— Peptic ulcer	Potential
— Acute pancreatitis	Confirmed
IV Inflammatory disorders	
— Rheumatic disease (rheumatoid arthritis)	Possible
V Cardiovascular disease	
— Migraine	Possible
— Immunoarteriosclerosis	Possible
— Raynaud's disease	Possible
— Other disorders of arterial circulation	Speculative
— Calcification of large vessels	Possible
— Arrhythmias	Speculative
— Senile cardiovacular disease	Speculative
VI Other conditions	
— Burns	Speculative
— Stress	Possible
— Psychiatric disorders	Speculative

Hypercalcemia and *hypercalcemic crisis* due to excessive osteolysis associated with cancer of the breast, lung or other organ, myeloma, hyperparathyroidism, immobilization, or vitamin D intoxication respond well to calcitonin in both short-term and long-term use.

Paget's disease (osteitis deformans) responds well, especially where there is bone pain with neurological complications, high bone turnover (as shown by raised serum alkaline phosphatase and urinary hydroxyproline levels) or progressive bone lesions with partial or repeated fractures.

Osteoporosis with high bone turnover due to estrogen deficiency or to primary or secondary hyperparathyroidism is also a classic indication for calcitonin. In the USA calcitonin is now accepted by the FDA as a treatment for postmenopausal osteoporosis.

Bone pain and *Sudeck's atrophy* (reflex sympathetic dystrophy) are well-established indications for calcitonin. It is also used in *acute pancreatitis* as an adjuvant to primary therapy.

Beneficial effects of calcitonin have also been reported in osteogenesis imperfecta [238, 239], in controlling bone loss during long-term treatment with steroids [240] or heparin, and in chronic renal insufficiency associated with excessive osteoclast activity [241].

11.1 Choice of Calcitonin

In Paget's disease SCT and HCT have comparable effects at equipotent hypocalcemic doses, and their efficacy is normally maintained for 3–20 months. In hypercalcemia, SCT and ECT are the calcitonins of choice. Overall, SCT and ECT are probably the best calcitonins for therapeutic use, followed by HCT and PCT.

11.2 Route of Administration and Patient Compliance

Extension of the use of calcitonin in clinical practice will depend on improvement in patient compliance and acceptability. An oral or other non-injectable preparation is essential if calcitonin is to realize its full therapeutic potential. Trials of an intranasal calcitonin spray suggest that this is likely to be an effective and acceptable alternative mode of administration.

11.3 Side-Effects

The principal side effects of calcitonin are shown in Table 10. Subcutaneous administration is tolerated better than intramuscular or intravenous, while intranasal administration is better than either. If impurities account for some side-effects, then HCT would be expected to cause more unwanted effects than SCT because larger quantities are needed.

11.4 Resistance

Resistance is sometimes a problem with calcitonin treatment, especially in long-term use. There are three forms of resistance — primary resistance (non-response), plateau phenomenon, and secondary (rebound) resistance.

In *primary resistance* the patient shows virtually no response to normal therapeutic doses and, often, still no response when the dose is raised, for example to 100–500 IU

Table 10. Adverse reactions occurring with calcitonins, especially after i.m. administration

Gastrointestinal symptoms
— Nausea
— Vomiting
— Abdominal pain
— Diarrhoea
— Unpleasant metallic taste in the mouth
Vascular symptoms
— Facial flushing
— Sensation of facial warmth
— Sensation of warmth affecting the hands
— Tingling in the extremities
Local symptoms
— Erythema at the site of injection
— Pain at the site of injection
Renal symptoms
— Increased frequency of micturition
— Polyuria
Allergic symptoms
— Rash

SCT/day [242] or 0.5–1 to 2.5–5 mg HCT/day [243]. Such patients, who are fortunately few, must simply be classed as non-responders.

In patients with Paget's disease exhibiting the *plateau phenomenon*, administration of calcitonin induces a fall in serum alkaline phosphatase, but the response is insufficient and blood levels cannot be brought below a certain point whatever the type and dosage of calcitonin given. Possible explanations include secondary hyperparathyroidism, the presence of antibodies to calcitonin, a reduction in the number of specific receptors, and the presence of calcitonin-resistant osteoclasts (which remain active) in addition to the normal calcitonin-sensitive osteoclasts, which are inhibited by the hormone.

Some aspects of the plateau phenomenon are poorly understood. In affected patients no further drop in alkaline phosphatase levels can be obtained even by withdrawing and then reinstituting calcitonin. Some patients who exhibit the phenomenon do not have antibodies to calcitonin in their blood. Furthermore, patients exhibiting the plateau phenomenon may nevertheless experience continued relief of symptoms, the histological findings may show steady improvement, and radiologically there may be continued progress from a focal negative to a focal positive bone balance [244–247].

In *secondary resistance* the inhibitory effect of calcitonin on bone resorption is initially normal but is then lost and the patient becomes unresponsive, sometimes with an exacerbation of the disorder for which calcitonin is being given. Some degree of secondary resistance is common, and in about 50% of cases it is associated with the formation of antibodies to calcitonin. (All forms of the hormone except natural HCT are potentially antigenic [248, 249]). To determine whether a patient is producing antibodies that interfere with the action of a particular calcitonin, the acute hypocalcemic response to two different calcitonins must be compared. Resumption of the response when a calcitonin of a different species is tried is the only proof that the resistance is mediated by anti-calcitonin antibodies [250]. In practice, the risk of antibody-mediated resistance might be reduced by using an intermittent treatment regimen, though in fact only 2% of patients have been found to develop antibody-related resistance.

12 Conclusions

Since the discovery of calcitonin by Copp in 1961, calcitonins have been identified in several animal species and in man. The individual types have been isolated and their structures elucidated, and some of them and one analogue have been synthesized. Four types are currently in therapeutic use, the two derived from fish, salmon and eel (aminosuberic analogue) calcitonins being considerably more potent than the mammalian varieties, human and porcine calcitonin.

All this has generated a vast body of literature on calcitonin which, however, contains many discrepant findings regarding both its physiological role and its pharmacological properties. Although many of these inconsistencies are probably more apparent than real, it is nevertheless prudent to continue to describe calcitonin conservatively as a hormone which, in combination with many other factors, has a "corrective and/or regulatory" effect on the following processes:

- Phosphorus and calcium metabolism under normal physiological conditions — for example the maintenance of calcium balance and skeletal mass. Calcitonin does this by controlling bone remodelling and the uptake, storage and elimination of calcium. It is also active in situations involving high calcium demand such as occur during bone growth, preganancy and lactation, and eating.
- Gastric and pancreatic function.
- Renal function, especially excretion of water and electrolytes.
- Certain endocrine functions, such as prolactin secretion.
- Certain neural mechanisms involved centrally in pain control and in neuromodulation.
- All biological processes in which the movement of calcium has a key role.

In therapeutic use, the calcitonins have demonstrated a low level, or even absence, of toxicity, with relatively minor side-effects such as facial flushing and nausea. They therefore offer a wide margin of safety. Their efficacy in diseases characterized by disorders of bone remodelling or bone loss, such as Paget's disease, some types of hypercalcemia, high-bone-turnover osteoporosis, and pain associated with osteoporosis, is well established, and they will very probably also prove useful in other diseases involving bone, such as periodontosis, as well as in non-skeletal diseases, especially diseases of the cardiovascular system (immunoarteriosclerosis, migraine, arrhythmias) and the nervous system.

A major prerequisite for such extension of the use and usefulness of calcitonin-based drugs, however, is the improvement of patient compliance and acceptability. Especially where long-term treatment is involved — as it inevitably is in many of these conditions — an oral or other non-injectable dosage form is essential if calcitonin is to realize its full therapeutic potential. One such development has recently emerged in the form of a spray solution for intranasal administration and of a suppository. Trials indicate that these new dosage forms are likely to offer a viable alternative to present forms of the drug.

13 References

1. Copp DH, Davidson AG (1961) Proc. Soc. Exp. Biol. Med. 107: 342
2. Copp DH et al. (1961) Proc. Canad. Fed. Biol. Soc. 4: 17
3. Foster GV, Baghdiantz A, Kumar MA, Slack E, Soliman HA, MacIntyre I (1964) Nature 202: 1303
4. Copp DH et al. (1962) Endocrinology 70: 638
5. Foster GV, MacIntyre I, Pearse AGE (1964) Nature 203: 1029
6. MacIntyre I et al. (1984) Biomed. Pharmacother. 38: 230
7. David L et al. (1975) Pathol. Biol. (Paris) 23: 833
8. Deftos LJ, O'Riordan JJH (1978) In: Copp DH, Talmage RV (eds) Proc. 6th Parathyroid Conf. Excerpta Medica Int. Congr. Ser. 421: 345
9. Straus E, Yalow RS (1976) J. Lab. Clin. Med. 87: 292
10. Deftos LJ (1983) In: Gennari C, Segre G (eds) The effects of calcitonins in man, Proc. 1st. Int. Workshop, Florence 1982. Masson, p 43
11. Fournie A et al. (1977) Rhum. 44: 91
12. Marx SJ, Woodward CJ, Auerbach GD (1972) Science 178: 999
13. Heersche JN, Marcus R, Auerbach GD (1974) Endocrinology 94: 241
14. Mahaffey JE et al. (1979) Calcif. Tiss. Int. 28: Suppl.: 154
15. Salmon DM et al. (1983) Mol. C. Endoc. 33: 293
16. Le Douarin N, Le Lièvre C (1970) C.R. Acad. Sci. Paris 270: 2857
17. Pearse AGE, Polak JM (1971) Histochemie 27: 96
18. Silva OL, Becker KL (1981) In: Pecile A (ed) Calcitonin 1980, Proc. Int. Symp, Milan 1980. Excerpta Medica Int. Congr. Ser. 540: 144
19. Milhaud G (1981) In: Pecile A (ed) Calcitonin 1980, Proc. Int. Symp., Milan 1980. Excerpta Medica Int. Congr. Ser. 540: 154
20. Kanis JA, Heynen G, Cundy T, Cornet F, Paterson A, Russel RG (1982) Clin. Sci. 65: 2
21. Cecchettin M, Tarquini B, Miolo M, Conte N (1986) Biomed. Pharmacother. 40: 19
22. Murad F, Brewer HB Jr, Vaughan M (1970) Proc. Natl. Acad. Sci. USA 65: 446
23. Marx SJ et al. (1972) Science 178: 999
24. Rao LG et al. (1981) Endocrinology 108: 1972
25. Chabardes D, Imbert Teboul M, Montegut M, Clique A, Morel F (1976) Proc. Natl. Acad. Sci. USA 73: 3608
26. Charbardes D, Gagnan Brunette M, Imbert Teboul M, Gontcharevskaia O, Montegut M, Clique A, Morel F (1980) J. Clin. Invest. 65: 439
27. Koida M, Nakamuta H, Furukawa S, Orlowski RC (1980) Jpn. J. Pharmacol. 30: 575
28. Fischer JA, Tobler PH, Kaufmann M, Born W, Henke H, Cooper PE, Sagar SM, Martin JB (1981) Proc. Natl. Acad. Sci. USA 78: 7801
29. Olgladl VR, Guidebone F, Nettl C, Pacile A (1983) Brain-Res. 265: 209
30. Henke H, Tobler PH, Fischer JA (1983) Brain-Res. 261: 373
31. Maurer R, Marbach P, Mousson R (1983) Brain-Res. 261: 346
32. Chausmer AB, Stevens MD, Severn C (1982) Science 216: 735
33. Marx SJ, Auerbach GD, Gavin JR, Buell DW (1974) J. Biol. Chem. 249: 6812
34. Moran J, Hunziker W, Fischer JA (1978) Proc. Natl. Acad. Sci. USA 75: 3984
35. Hunt NH, Ellison M, Underwood JC, Martin TJ (1977) Br. J. Cancer 35: 777

36. Findlay DM, Michelangeli VP, Eismann JA, Frampton RJ, Moseley JM, MacIntyre I, Whitehead R, Martin TJ (1980) Cancer Res. 40: 4764
37. Lamp SJ, Findlay DM, Moseley JM, Martin TJ (1981) J. Biol. Chem. 256: 12269
38. Findlay DM, Ng KW, Niall M, Martin TJ (1982) Biochem J. 206: 343
39. Heyne G, Franchimont P (1975) Pathol. Biol. (Paris) 23: 815
40. Parthemore JG, Deftos LJ, Bronzert D (1975) J. Clin. Invest. 56: 835
41. Roos BA, Parthemore JG, Lee JC, Deftos LJ (1977) Calcif. Tissue Res. 22: 298
42. Heyne G, Franchimont P (1974) Eur. J. Clin. Invest. 4: 213
43. Snider RH, Silva OL, Moor CF, Becker KL (1977) Clin. Chim. Acta. 76: 1
44. Austin LA, Heath H III (1981) N. Engl. J. Med. 304: 269
45. Feletti C et al. (1983) In: Gennari C, Segre G (eds) The effects of calcitonin in man, Proc. 1st. Int. Workshop, Florence 1982. 1982, p 249. Masson
46. Heath H III et al. (1984) Biomed. Pharmacother. 38: 241
47. Deftos LJ, Weisman MH, Williams GW, Karpf DB, Frumar AM, Davidson BJ, Parthemore JG, Judd HL (1980) N. Engl. J. Med. 302: 1351
48. Hillyard CJ, Cooke TJ, Coombes RC, Evans IM, MacIntyre I (1977) Clin. Endocrinol. (Oxf) 6: 291
49. Copp DH (1968) In: Talmage RV, Bélanger LF (eds) Proc. 3rd. Parathyroid Conf., Montreal 1967. Excerpta Medica Int. Cong. Ser. 159: 25
50. Potts JT Jr et al. (1968) In: Talmage RV, Bélanger LF (eds) Proc. 3rd. Parathyroid Conf., Montreal 1967. Excerpta Medica Int. Cong. Ser. 159: 407
51. Care AD et al. (1968) In: Talmage RV, Bélanger LF (eds) Proc. 3rd. Parathyroid Conf., Montreal 1967. Excerpta Medica Int. Cong. Ser. 159: 417
52. Sethi R et al. (1983) J. Clin. Endocrinol. Metab. 56: 549
53. Care AD et al. (1971) Endocrinology 89: 262
54. Talmage RV et al. (1981) In: Pecile A (ed) Calcitonin 1980, Proc. Int. Symp., Milan 1980. Excerpta Medica Int. Cong. Ser. 540: 96
55. Stevenson JC, Hillyard CJ, MacIntyre I, Cooper H, Whitehead MI (1979) Lancet 2: 769
56. Whitehead M, Lane G, Young O, Campbell S, Abeyasekera G, Hillyard CJ, MacIntyre I, Phang KG, Stenvenson JC (1981) Br. Med. J. (Clin. Res.) 283: 10
57. Munson PL et al. (1981) In: Pecile A (ed) Calcitonin 1980, Proc. Int. Symp., Milan 1980. Excerpta Medica Int. Cong. Ser. 540: 110
58. Toverud SU, Cooper CW, Munson PL (1978) Endocrinology 103: 472
59. Cannarozzi DB, Canale DD, Donabedian RK (1976) Clin. Chim. Acta 66: 387
60. Roos BA, Cooper CW, Frelinger AL, Deftos LJ (1978) Endocrinology 103: 2180
61. Stevenson JC, Evans IM (1981) Drugs 21: 257
62. Heynen G et al. (1981) In: Pecile A (ed) Calcitonin 1980, Proc. Int. Symp., Milan 1980. Excerpta Medica Int. Cong. Ser. 540: 208
63. Tiegs RD et al. (1986) J. Bone Min. Res. 1: 339
64. Dupuy B (1983) Biomed. Pharmacother. 37: 54
65. Fiore CE et al. (1981) In: Pecile A (ed) Calcitonin 1980, Proc. Int. Symp., Milan 1980. Excerpta Medica Int. Cong. Ser. 540: 170
66. Owyang C, Heath H III, Sizemore GW, Go VL (1978) Am. J. Dig. Dis. 23: 1084
67. Vora NM et al. (1978) J. Clin. Endocrinol. Metab. 46: 567
68. Leggate J, Farish E, Fletscher CD, McIntosh W, Hart DM, Sommerville JM (1984) Clin. Endocrinol. (Oxf) 20: 85
69. Stevenson JC, Abeyasekera G, Hillyard CJ, Phag KG, MacIntyre I, Campbell S, Townsend PT, Young O, Whitehead MI (1981) Lancet 1: 693
70. Christiansen C, Christensen MS, McNair P, Hagen C, Stocklund KE, Transbol I (1980) Eur. J. J. Clin. Invest. 10: 273
71. Recker RR, Saville PD, Heaney RP (1977) Ann. Intern. Med. 87: 649
72. Lindsay R, Hart DM, Aitken JM, MacDonald EB, Anderson JB (1976) Lancet 1: 1038
73. Stevenson JC, Whitehead MI (1982) Br. Med. J. 285: 585
74. Motta L, Maugeri D (1983) In: Gennari C, Segre G (eds) The effects of calcitonin in man, Proc. 1st. Int. Workshop, Florence 1982, Masson p 267
75. Taggart HM, Chesnut CH, Ivey JL, Baylink DJ, Sisom K, Huber MB, Roos BA (1982) Lancet 1: 475

76. Foresta C, Busnardo B, Ruzza G, Zanatta G, Mioni R (1983) Horm. Metab. Res. 15: 206
77. Body JJ, Demeester Mirkine N, Borkowski A, Suciu S, Corvilain J (1986) J. Clin. Endocrinol. Metab. 62: 700
78. Carey DE et al. (1980) J. Clin. Invest. 65: 892
79. Stevenson JC, Hillyard CJ (1980) Recent Results Cancer Res. 73: 60
80. Deftos LJ (1983) In: Gennari C, Segre G (eds) The effects of calcitonin in man, Proc. 1st. Int. Workshop, Florence 1982, Masson p 43
81. Weil C (1985) Klin. Wochenschr. 63: 433
82. Pearse AG, Polak JM (1971) Gut 12: 783
83. Hillyard CJ, Coombes RC, Greenberg PB, Galante LS, MacIntyre I (1976) Clin. Endocrinol. (Oxf) 5: 18
84. Avioli LV (1972) Kidney Int. 2: 241
85. Silva OL et al. (1974) Lancet 1: 1055
86. Lee JC, Parthemore JG, Deftos LJ (1977) J. Clin. Endocrinol. Metab. 45: 528
87. Heynen G et al. (1976) Lancet 2: 1322
88. Isaac R, Nivez MP, Piamba G, Fillastre JP, Ardaillou R (1975) Clin. Nephrol. 3: 14
89. Dambacher MA, Hunziker W, Fischer JA (1977) Dtsch. Med. Wochenschr. 102: 1191
90. Silva OL, Becker KL, Shalhoub RJ, Snider RH, Bivins LE, Moore CF (1977) Nephron 19: 12
91. Nielsen HE, Christensen CK, Olsen KJ (1979) Acta Med. Scand. 205: 615
92. Korman MG, Laver MC, Hansky J (1972) Br. Med. J. 1: 209
93. Ardaillou R, Beaufils M, Nivez MP, Isaac R, Mayaud C, Sraer JD (1975) Clin. Sci. Mol. Med., 49: 301
94. Ardaillou R, Sizonenko P, Meyrier A, Vallee G, Beaugas C (1970) J. Clin. Invest. 49: 2345
95. Canale DD, Donabedian RK (1975) J. Clin. Endocrinol. Metab. 40: 738
96. Avioli LV, Birge SJ, Scott S, Shieber W (1969) Am. J. Physiol. 216: 939
97. Robertson GM et al. (1976) New. Engl. J. Med. 294: 512
98. Condon JR, Ives D, Knight MJ, Day J (1975) Br. J. Surg. 62: 115
99. Norberg HP, DeRoos J, Kaplan EL (1975) Surgery 77: 773
100. Paloyan E et al. (1967) Surgery 62: 167
101. Lawrence AM (1972) In: Lefebre PJ, Unger RH (eds) Glucagon. Pergamon, p259
102. Weir GC, Lesser PB, Drop LJ, Fischer JE, Warshaw AL (1975) Ann. Intern. Med. 83: 185
103. Sowa M, Appert HE, Howard JM (1977) Surg. Gynecol. Obstet. 144: 365
104. Tagliaro F, Capra F, Dorizzi R, Luisetto G, Accoardini A, Renda E, Parolin A (1984) J. Endocrinol. Invest. 7: 331
105. Robinson MF et al. (1982) J. Clin. Endocrinol. Metab. 55: 538
106. Manolagas SC, Deftos LJ (1985) N. Engl. J. Med. 312: 122
107. Pavlinac DM, Lenhard LW, Parthemore JG, Deftos LJ (1980) J. Clin. Endocrinol. Metab. 50: 717
108. Arver S, Bucht E, Sjöberg HE (1984) Acta Physiol. Scand. 122: 461
109. Baron R, Saffar JL (1977) Calcif. Tissue Res. 22: 265
110. Frost HM (1964) In: Frost HM (ed) Bone dynamics. Little, Brown, p 315
111. Mundy CR, Altman AJ, Gondek MD, Bandelin JG (1977) Science 196: 1109
112. Teitelbaum SL, Stewart CC, Kahn AJ (1979) Calcif. Tissue Int. 27: 255
113. Chambers TJ, Dunn CJ (1983) Calcif. Tissue Int. 35: 566
114. Owen M (1980) Arthritis Rheum. 23: 1073
115. Owen M (1971) In: Bourne GH (ed) The biochemistry and physiology of bone. Academic, p271
116. Vaughan J (1981) Lancet 2: 133
117. Rodan GA, Rodan SB (1984) In: Peck WA (ed) Bone and mineral research. Elsevier, p244
118. Chambers TJ, Moore A (1983) J. Clin. Endocrinol. Metab. 57: 819
119. Martin TJ, Robinson CJ, MacIntyre I (1966) Lancet 1: 900
120. Accardo G et al. (1982) Curr. Ther. Res., Clin. Exp. 31: 422
121. Wase AW, Solewski J, Rickes F, Seidenberg J (1967) Nature 214: 388
122. McSheehy PMJ, Chambers TJ (1986) Endocrinology 119: 1654
123. Komm BS et al. (1987) J. Bone Min. Res. 2/Suppl. 1, abstract 237
124. Eriksen EF (1987) J. Bone Min. Res. 2/Suppl. 1, abstract 238
125. Eriksen EF (1987) In: Jensen J (ed) Abstracts for Int. Symp. on Osteoporosis, Aalborg 1987, abstract 67

126. Borle AB (1975) J. Membr. Biol. 21: 125
127. MacIntyre I, Parsons JA (1966) J. Physiol. Lond. 183: 31
128. Talmage RV et al. (1983) In: Peck WA (ed) Bone and mineral research. Excerpta Medica, p 74
129. Eilam Y, Szydel N, Harell A (1980) Mol. Cell. Endocrinol. 18: 215
130. Harell A, Binderman I, Guez M (1976) Isr. J. Med. Sci. 12: 115
131. Borle AB (1973) Fed. Proc. 32: 1944
132. Borle AB (1975) In: Talmage RV, Owen M, Parsons JA (eds) Calcium-regulating hormones. Excerpta Medica, Int. Cong. Ser. 346. p 217
133. Borle AB (1981) Rev. Physiol. Biochem. Pharmacol. 90: 13
134. DiPolo R (1979) J. Gen. Physiol. 73: 91
135. Denis G, Kuczerpa A (1974) Can. J. Physiol. Pharmacol. 52: 355
136. Raisz LG (1970) Fed. Proc. 29: 1176
137. Kawashima H, Torikai S, Kurokawa K (1981) Nature 291: 327
138. Caniggia A et al. (1981) In: Pecile A (ed) Calcitonin 1980, Proc. Int. Symp. Milan 1980. Excerpta Medica, Int. Cong. Ser. 540: 225
139. Haynes RC Jr, Murad F (1985) In: Gilman AG et al. (eds) The pharmacological basis of therapeutics. MacMillan, p 1517
140. Pearse AG (1966) Proc. R. Soc. Lond. 164: 478
141. Pearse AG, Carvalheira AF (1967) Nature 214: 929
142. Barreras RF (1973) Gastroenterology 64: 1168
143. Case RM (1973) Digestion 8: 269
144. Grodsky GM, Bennett LL (1966) Diabetes 15: 910
145. Deftos LJ, Burton D, Bone HG, Catherwood BD, Parthemore JG, Moore RY, Minick S, Guillemin R (1978) Lift Sci. 23: 743
146. Flynn JJ, Margules DL, Cooper CW (1981) Brain Res. Bull. 6: 547
147. Nicoletti F, Clementi G, Patti F, Canonico PL, Di. Giorgio RM, Matera M, Pennisi G, Angelucci L, Scapagnini U (1982) Brain Res. 250: 381
148. Rizzo AJ, Goltzman D (1981) Endocrinology 108: 1672
149. Segre G et al. (1983) In: Gennari C (ed) The effects of calcitonin in man, Proc. 1st. Int. Workshop, Florence 1982. Masson, p 55
150. Deftos LG, Parthmore JG (1981) In: Pecile A (ed) Calcitonin 1980, Proc. Int. Symp., Milan 1980. Excerpta Medica Int. Cong. Ser. 540: 136
151. Copp DH (1983) In: Gennari C, Segre G (eds) The effects of calcitonin in man, Proc. 1st. Int. Workshop, 1982, Florence 1982. Masson, p 3
152. Peck WA (1979) Calcif. Tissue Int. 29: 1
153. Russel RG et al. (1981) Metab. Bone Dis. and Rel. Res. 4/5: 255
154. Varghese Z, Moorhead JF, Wills MR (1981) Clin. Chim. Acta 110: 105
155. Delmas PD, Stenner D, Wahner HW, Mann KG, Riggs BL (1983) J. Clin. Invest. 71: 1316
156. Slovik DM, Gundberg CM, Neer RM, Lian JB (1984) J. Clin. Endocrinol. Metab. 59: 228
157. Price PA et al. (1979) Calcif. Tissue Int. 28: 159
158. Deftos LJ, Parthemore JG, Price PA (1982) Calcif. Tissue Int. 34: 121
159. Menon RK et al. (1983) J. Clin. Endocrinol. Metab. 1: 307
160. Epstein S, Poser J, McClintock R, Johnston CC Jr, Bryce G, Hui S (1984) Lancet 1: 307
161. Delmas PD, Wilson DM, Mann KG, Riggs BL (1983) J. Clin. Endocrinol. Metab. 57: 1028
162. Chambers TJ, Dunn CJ (1983) Calcif. Tissue Int. 35: 566
163. Heersche JNM et al. (1981) In: Pecile A (ed) Calcitonin 1980, Proc. Int. Symp., Milan 1980. Excerpta Medica Int. Cing. Ser. 540: 67
164. Haas HG, Dambacher MA, Cuncaga J, Lauffenbruger T (1971) J. Clin. Invest. 50: 2689
165. Ardaillou R, Fillastre JP, Milhaud G, Rousselet F, Delaunay F, Richet G (1969) Proc. Soc. Exp. Biol. Med. 131: 56
166. Ardaillou R, Milhaud G, Rousselet F, Vuagnat P (1967) C. R. Acad. Sci. (D) (Paris) 264: 3037
167. Langer B, Peytremann A, Rufener C, Jenny M (1971) Schweiz. Med. Wochenschr. 101: 69
168. Ardaillou R, Vuagnat P, Milhaud G, Richet G (1967) Nephron 4: 298
169. Loreau N, Lepreux C, Ardaillou R (1975) Biochem. J. 150: 305
170. Murad F, Brewer HB Jr, Vaughan M (1970) Proc. Natl. Acad. Sci. USA 65: 446
171. Heersche JN et al. (1974) Endocrinology 94: 241
172. Agus ZS et al. (1981) Ann. Rev. Physiol. 43: 585

173. Puschett JB, Beck WS Jr, Jelonek A, Fernandez PC (1974) J. Clin. Invest. 53: 756
174. Williams CP, Meachim G, Taylor WH (1978) J. Clin. Pathol. 31: 1212
175. Gray TK, Brannan P, Juan D, Morawski G, Fordtran JS (1976) Gastroenterology 71: 392
176. Kisloff B, Moore EW (1977) Gastroenterology 72: 462
177. Walling MW, Brasitus TA, Kimberg DV (1977) Gastroenterology 73: 89
178. Gennari C et al. (1983) In: Gennari C, Segre G (eds) The effects of calcitonin in man, Proc. 1st. Int. Workshop, Florence 1982, Masson p 93
179. Bernier JJ, Rambaud JC, Cattan D, Prost A (1969) Gut 10: 980
180. Goebell H, Ammann R, Herfarth C, Horn J, Hotz J, Knoblauch M, Schmid M, Jaeger M, Akovbiantz A, Linder E, Abt K, Nuesch E, Barth E (1979) Scand. J. Gastroenterol. 14: 881
181. Goebell H, Hotz J (1981) In: Pecile A (ed) Calcitonin 1980, Proc. Int. Symp., Milan 1980. Excerpta Medica Int. Cong. Ser. 540: 346
182. Hotz J et al. (1978) Therapiewoche 28: 1066
183. Luisetto G et al. (1981) In: Pecile A (ed) Calcitonin 1980, Proc. Int. Symp., Milan 1980. Excerpta Medica Int. Cong. Ser. 540: 336
184. Paul F, Ohnhaus EE, Hesch RD, Chemnitz G, Hope Seyler R, Henrichs HR, Hartung H, Waldmann D, Kunze K, Barth E, Nuesch E, Abt K (1979) Dtsch. Med. Wochenschr. 104: 615
185. Maier R (1977) In: MacIntyre I (ed) Human calcitonin and Paget's disease, Proc. Int. Workshop, London 1976, Huber p 66
186. Azria M, Kiger JL (1974) Thérapie 29: 753
187. Ziliotto D et al. (1976) Minerva Endocrinol. 1: 159
188. Milhaud G (1978) In: Giroud JP et al. (eds) Pharmacologie clinique, vol 1; p 859. Expansion scientifique française, Paris.
189. Pecile A et al. (1983) In: Gennari C, Segre G (eds) The effects of calcitonin in man, Proc. 1st. Int. Workshop, Florence 1982, p 205. Masson
190. Kleibel F et al. (1983) Neurosci. Lelt. Suppl. 14: 199
191. Allan E (1983) Pharmatherapeutica 3: 482
192. Gennari C et al. (1981) Min. Metab. Res. It. 2: 109
193. Gennari C et al. (1983) In: Gennari C, Segre G (eds) The effects of calcitonin in man, Proc. 1st. Int. Workshop, Florence 1982, p 213. Masson
194. Welzel D (1983) In: Gennari C, Segre G (eds) The effects of calcitonin in man, Proc. 1st. Int. Workshop, Florence 1982, p 223. Masson
195. Fischer JA, Sagar SM, Martin JB (1981) Life Sci. 29: 663
196. Laurian L et al. (1986) Horm. Metab. Res. 18: 168
197. Abdullahi SE, De-Bastiani G, Nogarin L, Velo GP (1975) Agents Actions 5: 371
198. Fiore CE et al. (1983) In: Gennari C, Segre G (eds) The effects of calcitonin in man, Proc. 1st. Int. Workshop, Florence 1982, p 291. Masson
199. Kawashima H, Torikai S, Kurokawa K (1981) Nature 291: 327
200. MacIntyre I (1983) Triangle (The Sandoz Journal of Medical Science) 22: 69
201. Hagemann J et al. (1977) Acta Endocrinol. Suppl. 212: 257
202. Ross AJ, Cooper CW, Ramp WK, Wells SA Jr (1980) Proc. Soc. Exp. Biol. Med. 163: 315
203. Chapuy MC, David L, Meunier PJ (1980) Horm. Metab. Res. 12: 486
204. Hill CS Jr, Ibanez ML, Samann NA, Ahearn MJ, Clark RL (1973) Medicine (Baltimore) 52: 141
205. Chabardes D, Imbert-Teboul M, Montegut M, Clique A, Morel F (1976) Proc. Natl. Acad. Sci. USA 73: 3608
206. Calabro A et al. (1984) Boll. Soc. Ital. Biol. Sper. 60: 1145
207. Guistina G et al. (1981) In: Pecile A (ed) Calcitonin 1980, Proc. Int. Symp., Milan 1980. Excerpta Medica Int. Cong. Ser. 540: 326
208. Petralito A, Lunetta M, Liuzzo A, Fiore CE, Heynen G (1979) J. Endocrinol. Invest. 2: 209
209. Sgambato S et al. (1978) Il Farmaco 33: 256
210. Fahrenkrug J et al. (1975) J. Clin. Endocrinol Metab. 41: 149
211. Decker HD et al. (1974) Am. J. Surg 127: 71
212. Leicht E, Biro G, Weinges KF (1974) Horm. Metab. Res. 6: 410
213. Isaac R et al. (1980) J. Clin. Endocrinol. 50: 1011
214. Cantalamessa L, Cantania A, Reschini E, Peracchi M (1978) Metabolism 27: 987

215. Pecile A et al. (1981) In: Pecile A (ed) Calcitonin 1980, Proc. Int. Symp., Milan 1980. Excerpta Medica Int. Cong. Ser. 540: 183
216. Ziliotto D et al. (1981) Horm. Metab. Res. 13: 64
217. Chihara K, Iwasaki J, Iwasaki Y, Minamitani N, Kaji H, Fujita T (1982) Brain Res. 248: 331
218. Riesterer L, Jaques R (1970) Pharmacology 3: 53
219. Ceserani R, Colombo M, Olgiati VR, Pecile A (1979) Life Sci. 25: 1851
220. Robert L et al. (1978) Ann. Biol. Anim. Bioch. Biophys. 18: 195
221. Robert L, Brechemier D, Godeau G, Labat ML, Milhaud G (1977) Biochem. Pharmacol. 26: 2129
222. Dupuy B (1983) Biomed. Pharmacother. 37: 54
223. Morley JE, Levine AS, Silvis SE (1981) Science 214: 671
224. Beveridge T, Niederer W, Nuesch E, Petrin A (1976) Z. Gastroent (Verh) 10: 12
226. Habener JF et al (1971) In: Talmage RV, Munson PL (eds) Calcium, parathyroid hormone and the calcitonins, Proc. 4th parathyroid Conf., Chapel Hill (NC) 1971. Excerpta Medica Int. Cong. Ser. 243: 152
227. Newsome FE et al. (1973) Endocrinology 92: 1102
228. Milhaud G, Hankiss J (1969) C.R. Acad. Sci. (D) (Paris) 268: 124
229. Fillastre JP et al. (1975) Europ. J. Intensive Care Med. 1: 185
230. Hosking DJ (1980) Metab. Bone Dis. and Rel. Res. 2: 207
231. Singer FR et al. (1978) In: Avioli LV, Krane SM (eds) Metabolic bone disease II, p 489 Academic
232. Agrawal R et al. (1981) In: Pecile A (ed) Calcitonin 1980, Proc. Int. Symp., Milan 1980. Excerpta Medica Int. Cong. Ser. 540: 237
233. Chesnut CH III et al. (1981) In: Pecile A (ed) Calcitonin 1980, Proc. Int. Symp., Milan 1980. Excerpta Medica Int. Cong. Ser. 540: 247
234. Vigo P et al. (1981) In: Pecile A (ed) Calcitonin 1980, Proc. Int. Symp., Milan 1980. Excerpta Medica Int. Cong. Ser. 540: 256
235. Wallach S et al. (1977) Curr. Ther. Res. 22: 556
236. Gennari C (1981) In: Pecile A (ed) Calcitonin 1980, Proc. Int. Symp., Milan 1980. Excerpta Medica Int. Cong. Ser. 540: 277
237. Goebell H, Hotz J (1981) In: Pecile A (ed) Calcitonin 1980, Proc. Int. Symp., Milan 1980. Excerpta Medica Int. Cong. Ser. 540: 346
238. Castells S et al. (1979) J. Pediat. 95: 807
239. Rosenberg E, Lang R, Boisseau V, Rojanasathit S, Avioli LV (1977) J. Clin. Endocrinol. Metab. 44: 346
240. Hahn TJ, Boisseau VC, Avioli LV (1974) J. Clin. Endocrinol. Metab. 39: 274
241. Avioli LV, Gennari C (1983) In: Gennari C, Segre G (eds) The effects of calcitonin in man, Proc. 1st. Int. Workshop, Florence 1982, p 103. Masson
242. Hamilton CR Jr (1974) Am. J. Med. 56: 315
243. Evans IMA (1977) In: MacIntyre I (ed) Human calcitonin and Paget's disease, Proc. Int. Workshop, London 1976, p 111. Huber
244. Nagant de Deuxchaisnes C et al. (1977) In: MacIntyre I, Szelke M (eds) Molecular Endocrinology, p 213. Elsevier
245. Nagant de Deuxchaisnes C et al. (1979) In: MacIntyre I, Szelke M (eds) Molecular Endocrinology, p 405. Elsevier
246. Nagant de Deuxchaisnes C et al. (1981) Acta Rhumatol. 4: 425
247. Nagant de Deuxchaisnes et al. (1980) Arthritis Rheum. 23: 1215
248. Singer FR, Aldred JP, Neer RM, Krane SM, Potts JT Jr, Bloch KJ (1972) J. Clin. Invest. 51: 2331
249. Woodhouse NJ, Mohamedally SM, Saed Nejad F, Martin TJ (1977) Br. Med. J. 2: 927
250. Martin TJ (1979) Aust. NZ. J. Med. 9: 36

Mafosfamide — A Derivative of 4-Hydroxycyclophosphamide

Ulf Niemeyer, Jürgen Engel, Peter Hilgard, Manfred Peukert, Jörg Pohl and Herbert Sindermann

ASTA Pharma AG, P.O. Box 140129, 4800 Bielefeld 14, FRG

The chemical development of mafosfamide (MA)-L-lysine salt as a "stabilized" prodrug of 4-hydroxy-CP, i.e. the solid-state unstable primary metabolite of the anticancer drug cyclophosphamide (CP), involved a stereoselective synthesis and an optimized cation exchange with subsequent lyophilization. The stereochemical and physicochemical characterization of MA elucidated further the complex metabolic behavior of oxazaphosphorine cytostatics and confirmed imino-CP as a transient primary intermediate.

Biologically, MA opens up new possibilities in clinical therapy and in preclinical tests, e.g. in the clonogenic stem cell assay, in in-vitro purging prior to autologous bone marrow transplantation (ABMT), and in the locoregional therapy of tumors. A completely new type of therapy is emerging, specifically for low-dose MA, to be used as an immunomodulator, possibly in combination with further substances from the biological response modifier group.

1 Introduction

Alkylating agents are the most frequently used drugs for the chemotherapy of cancer besides hormones and antibiotics. Their cytotoxic effect is caused by the alkylation of nucleophilic centres of biomolecules, e.g. DNA and proteins, inhibiting proliferation and growth and leading to cell death. However, alkylants are not only responsible for the destruction of tumor cells, but they are also detrimental to normal tissue, in particular to quickly proliferating cells. Therefore, great efforts are made to obtain compounds with a higher selectivity by means of chemical modification.

The oxazaphosphorines constitute the therapeutically most important drugs among alkylants. Their alkylating 2-chloroethylamine groups are bound to a 2-oxo-tetrahydro-1,3,2-oxazaphosphorine ring system which decisively enhances the selectivity of the drug's activity. In the mid 1950s this compound class was discovered by Arnold, Bourseaux and Brock in the research laboratories of ASTA Pharma and subsequently developed further in order to gain an even higher cancerotoxic selectivity [1].

The first compound was cyclophosphamide (CP, Endoxan) which was made available to clinicians in 1958, followed by trofosfamide (Ixoten) and ifosfamide (Holoxan) in the 1970s (Fig. 1). Due to the broad spectrum of clinical activity and the wide therapeutic range CP was considered by the World Health Organization (WHO) as standard drug for the evaluation of newly developed cytostatics (Geneva, 1974). Moreover, the worldwide scientific interest focused on the fact that oxazaphosphorines do not act per se as alkylants but — in vivo — undergo a fascinating metabolization cascade which can only be explained with great difficulty. Nowadays many working groups are increasingly engaged in the identification of metabolites

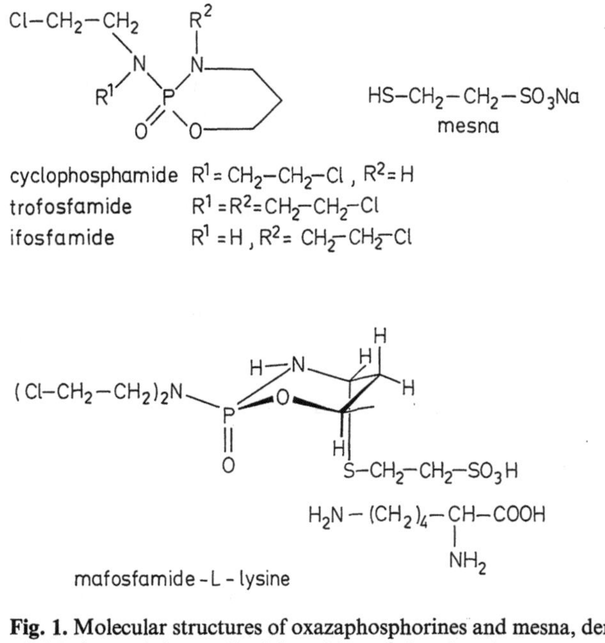

Fig. 1. Molecular structures of oxazaphosphorines and mesna, demonstrating the chair-conformation of the ring and the *cis*-diaxial position of the phosphorus oxygen and the 4-thio-substituent in MA

and the mechanisms of action, in order to gain an insight into the biochemical peculiarities of the cancer cells, and to combat them with a drug development which is as rational as possible.

For many years numerous hypotheses have competed in this respect until finally Brock and Hohorst detected the preferred metabolization of CP by liver microsomes [2], suggesting for the first time the structure of the highly unstable primary metabolite [3] and its key role in cancerotoxic selectivity [4]. This enlightenment modified also the former hypothesis of a selective splitting of CP in tumor cells by acid phosphatases to bis(2-chloroethyl)-amine. After comprehensive investigations into the mechanism of action of oxazaphosphorines Brock et al. finally investigated the cause of urotoxicity of oxazaphosphorines [5], introducing subsequently an optimal antidote into clinical therapy, namely sodium-2-mercaptoethanesulfonate (INN: mesna), Uromitexan (Fig. 1) [6]. The detoxifying effect of mesna is based on the stabilization of the primary metabolite of oxazaphosphorines in the urine, due to conjugation of both. The chemically synthesized CP-mesna conjugate (INN: Mafosfamide, MA) was then evaluated as a prodrug of the active primary metabolite of CP and of the uroprotective mesna. At present MA is clinically in phase I studies as L-lysine salt (Fig. 1).

2 Development of MA

2.1 Biotransformation of CP

The biotransformation is similar for all oxazaphosphorine cytostatics. The pathway to directly alkylating agents is shown for CP in Fig. 2. The initial step is an hydroxylation by hepatic microsomes at carbon atom 4 of the oxazaphosphorine ring. The activation product 4-hydroxy-CP degrades mainly spontaneously into phosphoramide mustard (*N*,*N*-bis(2-chloroethyl)-phosphorodiamidic acid) and acrolein via the ring opened tautomer aldophosphamide. In 1984, in vitro ^{1}H- and ^{13}C-NMR studies provided the first definitive evidence for the pseudo-equilibrium mixture of 4-hydroxy-CP consisting of *cis*- and *trans*-4-hydroxy-CP, aldophosphamide and its hydrate in the approximate ratio of 4:3:0.3:1 [7]. Moreover, the β-elimination of phosphoramide mustard from aldophosphamide was confirmed as rate limiting, first suggested by Hohorst et al. [8].

Under release of chloride, phosphoramide mustard undergoes intramolecular alkylation to a phosphoric aziridinium ion intermediate, which can be considered as the ultimate alkylans of CP [9]. This intermediate cannot be isolated, but it is proven by deuterium labeling experiments [10].

After a first alkylation, the second 2-chloroethyl-group reacts with an additional base of the DNA-helix, leading to cytotoxic interstrand or intrastrand cross-links [11]. Although there is in general a good correlation between DNA-cross-linking and growth inhibition, additional modes of action are possible. For instance, alkylating agents cause modifications of the plasma membrane and thereby inhibit cell proliferation [12].

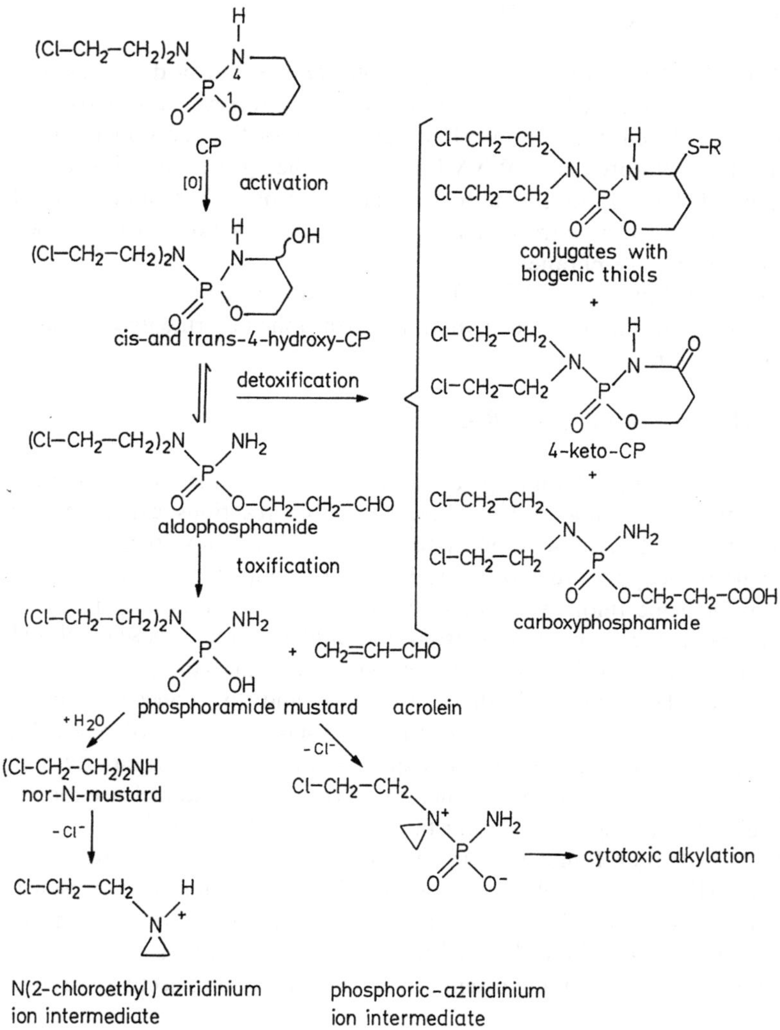

Fig. 2. Biotransformation of cyclophosphamide

Other direct alkylating intermediates, besides the phosphoric aziridinium ion, do not play a significant role in the cytotoxic activity of CP. Especially, a cytostatic activity of the aldehyde group of aldophosphamide and of acrolein was excluded [13]. Furthermore, the hydrolysis of phosphoramide mustard to Nor-*N*-mustard (Bis(2-chloroethyl)amine), transformed into the active alkylant *N*(2-chloroethyl)aziridinium ion, is negligible (Fig. 2) [14].

4-Hydroxy-CP plays a key role in the bioactivity, especially the cytotoxic specificity and cancerotoxic selectivity, of CP [4]. The rate and the position of the release of phosphoramide mustard from 4-hydroxy-CP seems to be the most important feature for specific alkylation.

2.2 Key Role of 4-Hydroxy-CP

4-Hydroxy-CP is the carrier-molecule of phosphoramide mustard across cell membranes. It penetrates mouse Ehrlich Ascites tumor cells in contrast to phosphoramide mustard [15,16]. Recently the intra-cellular chemistry of the active metabolites of CP were directly monitored by ^{31}P-NMR-studies utilizing a human histiocytic lymphoma cell line (U 937). Intracellular concentrations of phosphoramide mustard could only be attributed to the intracellular fragmentation of 4-hydroxy-CP, which readily permeates the cell membrane [17].

The degradation of 4-hydroxy-CP is influenced by various mechanisms. In vitro 4-hydroxy-CP reacts with different nucleophiles to oxazaphosphorine-intermediates according to the equation

$$\text{4-OH-CP} + \text{H-XR} \rightleftharpoons \text{4-XR-CP} + H_2O$$

or to acyclic structures like aldophosphamide oximes [18], hydrazones [19], imines, oxazolidones or thiohemiacetales [20,21]. These reversible reactions cause varying degrees of retardation in the release of phosphoramide mustard. In vivo analog reactions with biogene nucleophiles are to be expected. The temporary stabilization of 4-hydroxy-CP by biogenic thiols was studied by the group of Hohorst (Fig. 2) [22].

On the other hand, the β-elimination from aldophosphamide is base-catalyzed and can be intracellularly accelerated by the general base catalytic activity which varies directly with pH, ionic strength, temperature and the concentration of the base [23]. Moreover, Hohorst et al. found, that the P-O-bond of 4-hydroxy-CP is cleaved by various 3′,5′-exonucleases and phosphordiesterases. A particularly high activity shows the Kornberg-DNA-Polymerase with a 3′,5′-exonuclease-moiety catalyzing the release of phosphoramide mustard quite near the DNA [24].

Additionally, 4-hydroxy-CP underlies an enzymatic reaction to carboxyphosphamide and keto-CP (Fig. 2), both of which cannot release phosphoramide mustard. This detoxification was proposed to be different in normal and tumor cells and to give a rise of selectivity [25]. But, in the light of more recent investigations, this detoxification apparently does not contribute essentially to the oncostatic selectivity in vivo of CP [13].

The curative activity of 4-hydroxy-CP is higher than for all other metabolites [1]. The pharmacological characterization bases firstly on the cytostatic specificity in vitro, which is the quotient of the alkylating activity measured by 4(4-nitrobenzyl) pyridine (NBP-assay), and the cytotoxic activity determined in the incubation (transplantation test), and secondly on the cancerotoxic selectivity in vivo determined by the dose ratio between cancerotoxic action and systemic lethal action. In this respect 4-hydroxy-CP is clearly superior to all other nitrogen mustard compounds.

2.3 Urotoxicity of CP

In clinical use the oxazaphosphorine cytostatics cause urotoxic side-effects which manifest mainly as hemorrhagic cystitis. In the case of CP only the renally eliminated 4-hydroxy metabolite is responsible for these side-effects [5]. Searching for an antidote by Brock et al. [5,6,26,27] led to the development of mesna (sodium 2-mercaptoethanesulfonate), which by concomitant administration with CP reliably prevents urotoxicity. Mesna is oxidized in the blood into its disulfide (INN: dimesna) which is eliminated via the kidneys. Dimesna is partly reduced to mesna by interaction with the intracellular glutathione system of epithelial cells from the renal tubulus [28]. Therefore, 4-hydroxy-CP can be detoxified specifically in the bladder by the reaction with mesna. We found after CP/mesna-treatment in the urine a spontaneously formed mixture of MA and its *trans*-epimer by thin-layer chromatography with subsequent phosphorus-test or NBP-test. If there is an excess of mesna, the concentration of 4-hydroxy-CP is drastically reduced due to the law of mass action and the bladder is protected. With mesna in the urine the spontaneous degradation of 4-hydroxy-CP into the ten times more urotoxic acrolein and the less toxic phosphoramide mustard is stopped. In this case, a possible detoxification of acrolein [29] by reaction to acrolein-mesna-adducts should play no important role in the urine.

The development of mesna as a regional uroprotector has led to a further increae in cancerotoxic selectivity of the oxazaphosphorine cytostatics. The dose regimen can be enhanced, optimizing the therapeutic efficacy and possibly overcoming the occurrence of primary or secondary resistance [30].

3 Chemistry of MA

3.1 Synthesis

The key role of 4-hydroxy-CP for the bioactivity of CP was suggestive to initiate clinical studies. But this compound is difficult to prepare and not stable enough for drug development. The chemical approaches for a stabilization involved the synthesis of derivatives of 4-hydroxy-CP, which are stable in the solid form and which will spontaneously hydrolyze to 4-hydroxy-CP under physiological conditions. Hohorst et al. investigating the temporary stabilization of 4-hydroxy-CP in vivo by biogenic thiols, studied in vitro the reactions of 4-hydroxy-CP with different low molecular thiols. For instance, the reaction with 2-mercaptoethanol resulted in 4(2-hydroxyethyl)thio-CP [31]. Another CP-thiol conjugate was 4(2-carboxyethyl)thio-CP, resulting from a reaction with 3-mercaptopropionic acid [32].

The low toxicity of mesna (sodium 2-mercaptoethane sulfonate), the good water solubility and the uroprotective activity makes mesna a preferential candidate for conjugation with the primary metabolite of CP. A suitable synthesis of the CP-mesna conjugate was developed (Fig. 3). A critical step is the preparation of the unstable 4-hydroxy-CP, which was first published by Takamizawa et al. [33]. The synthetic strategy involved the ozonization of 3-butenyl-*N,N*-bis(2-chloroethyl)phosphorodiamidate yielding after addition of hydrogen peroxide the good crystallizing *cis*-epimer of 4-hydroperoxy-CP, which is stable at —25 °C. An alternative approach for 4-hydroperoxy-CP is the direct ozonization of the readily available CP [34]. 4-Hydroperoxy-CP is reduced by triphenylphosphine or more conveniently by aqueous sodium thiosulfate to *cis/trans*-4-hydroxy-CP. The subsequent condensation with mesna

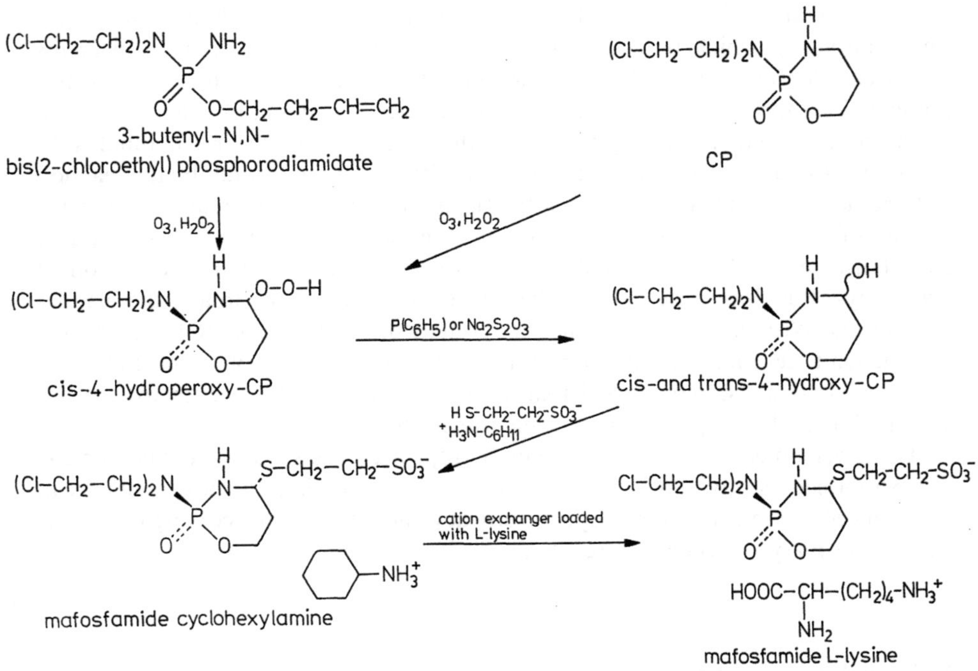

Fig. 3. Synthesis scheme for MA-L-lysine salt (only one antipode of the racemic *cis*-isomer is shown)

to MA-sodium salt proved to be difficult due to problems of isomerisation and isolation. In contrast to this, cyclohexylammonium 2-mercaptoethanesulfonate reacted stereoselectively in high yield in aqueous acetone to the crystallizable racemic *cis*-isomer [35]. In the last reaction step cyclohexylammonium ion is exchanged by L-lysinate using a strong acid cation exchange resin [36]. Regarding the rapid hydrolysis of MA, the cation exchange has to take place at strongly defined conditions. The eluted aqueous solution is adjusted at pH 4 and lyophilisated to a white dry residue of MA L-lysine salt.

3.2 Stereochemistry

Both the carbon atom 4 and the phosphorus atom are chiral centers. Therefore, two racemic diastereomers of C-4-substituted oxazaphosphorines are to be expected. The relative configuration of MA as *cis*-isomer was proven by single crystal X-ray diffraction of its cyclohexylamine salt (Fig. 4). The oxazaphosphorine ring adopts roughly a chair conformation. The phosphoryl oxygen and the 2-sulfonatoethylthio-side chain are situated approximately *cis*-diaxially to the heterocyclus. Consequently, the bis(2-chloroethyl)amino-group is in an equatorial position. It is worth mentioning that only one antipode was found in the crystal lattice. This separation of the MA antipodes by crystallization is a specific phenomenon in the oxazaphosphorine field [36]. All solutions of MA-cyclohexylamine showed no optical activity polarimetrically.

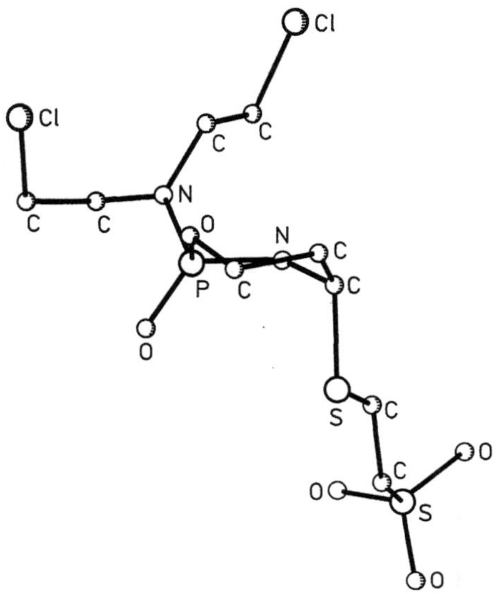

Fig. 4. Crystal structure of MA

The observable optical rotation of MA-L-lysine is only attributed to the protonated (+)-L-lysine. Therefore, dissolved MA is a racemic anion.

The stereochemistry of MA is comparable in solution and in the crystalline form. The roughly axial position of the C-4-substituent of dissolved MA was confirmed by the coupling constants of the proton at C-4 [35]. By high-field NMR using 2D experiments, including the phosphorus-decoupled H,H-COSY-technique, all questions concerning spin-assignment and molecular structure in solution state could be answered. Low and high-temperature NMR showed clearly that only one conformation is preferred in solution [36]. In contrast to this, different proportions of conformations had been reported for CP, *cis*-4-methyl-CP and some other oxazaphosphorines [37,38]. Referring to the diaxial patterns of the phosphoryl oxygen and the C-4-substituent MA and *cis*-4-hydroxy-CP possess the same stereochemistry.

It is of interest that the biological activity of 4-hydroxy-CP is suggested to be linked with the ring chair conformation which shows a strong similarity in the bond distances and angles to the cyclic phosphate moiety of cyclic adenosine 3′,5′-monophosphate (cAMP) and should be favorable for the ability to compete with cAMP for specific protein binding [39].

3.3 Behavior in Aqueous Solution at pH 7

In aqueous solution MA L-lysine is fully dissociated into the sulfonate-anion and the protonated L-lysine which shows in the range of pH 4 to 7 the typical zwitter ion structure of an amino acid. The solubility in water is over 50%. In vitro at pH 7 epimerization and hydrolysis start immediately after dissolution.

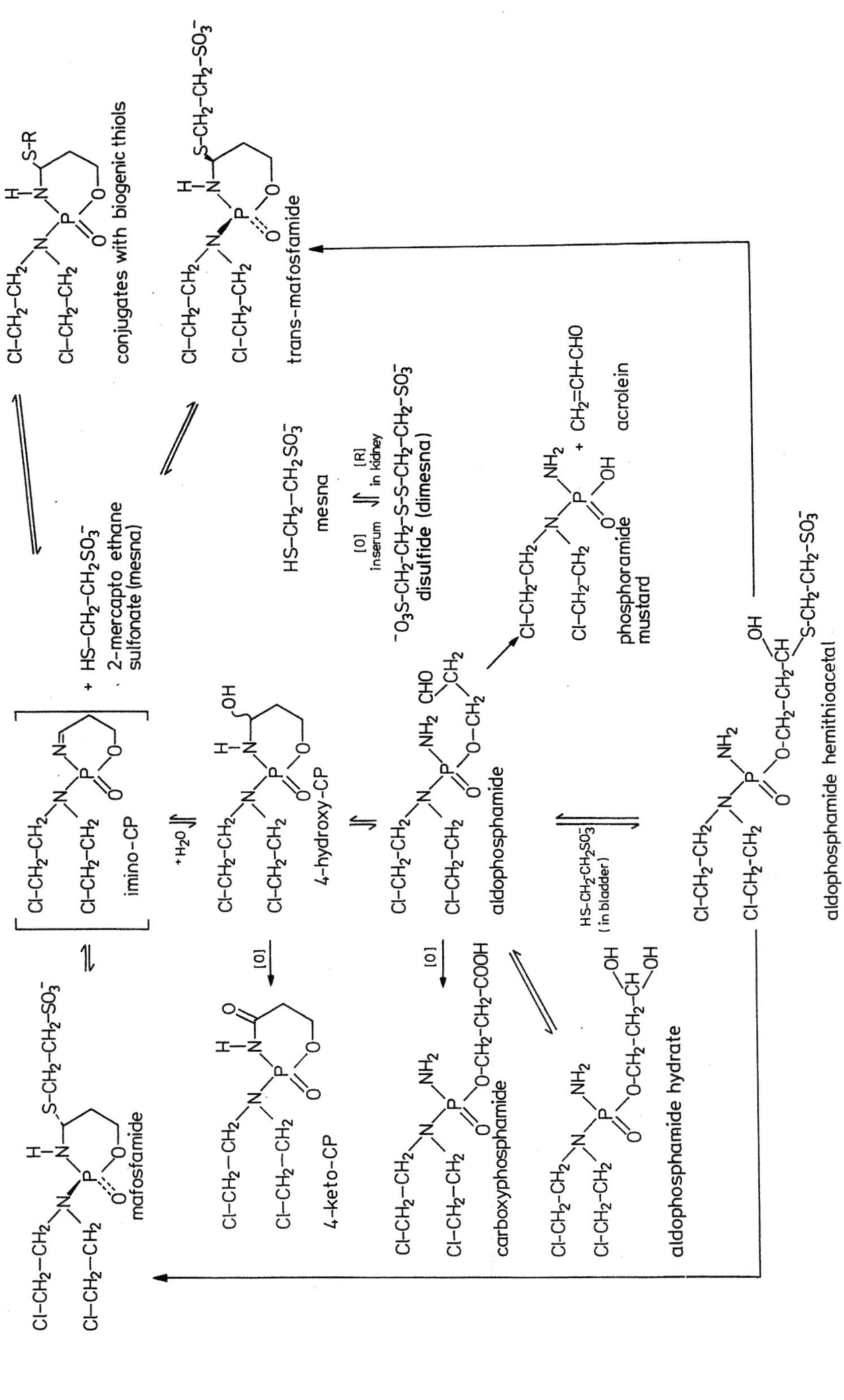

Fig. 5. Activation and biotransformation of MA (*cis*-4-sulfoethylthio-CP)

Epimerization

MA isomerizes into the *trans*-epimer (Fig. 5). The time course for the single isomers measured by HPLC is shown in Fig. 6. Within less than 5 minutes the *cis-trans*-equilibrium of 59 to 41 has been adjusted. This ratio remains constant during hydrolytic degradation of MA as indicated by parallel curves [35].

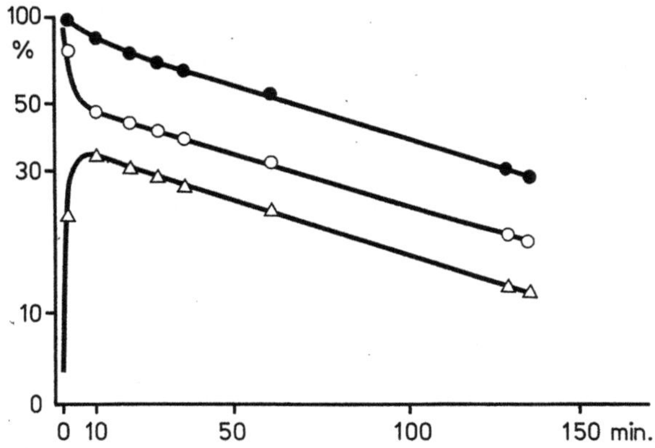

Fig. 6. Epimerization and hydrolysis of MA. Time courses for MA plus *trans*-epimer (●), MA (○) and *trans*-epimer (△) in 0.07 M phosphate buffer, pH 7.0 at 37 °C, initially containing 0.02 M of MA. Procedure: HPLC, C_{18} nucleosil; detection at 208 nm

Hydrolysis

MA hydrolyzes reversibly into 4-hydroxy-CP and 2-mercaptoethane sulfonate. Within a few minutes a "pseudo-equilibrium" is adjusted which is disturbed by an irreversible degradation of 4-hydroxy-CP via aldophosphamide to phosphoramide mustard and acrolein (Fig. 6). The initial rate of hydrolysis is more rapid than the final rate. The equilibrium constant of the reaction (mafosfamide + $H_2O \rightleftharpoons$ 4-hydroxy-CP + mesna) was determined as 5×10^{-4} mol/l at 37 °C in 0.07 M phosphate buffer of pH 7.0 [40]. Figure 7 shows the kinetic of 4-hydroxy-CP, mesna, phosphoramide mustard and acrolein-mesna-adducts within the first 160 minutes of the degradation of MA.

Starting with the *trans*-epimer instead of MA, epimerization and hydrolysis occur in an analog manner. This indicates that both isomers possess the same chemical reactivity.

In vitro the hydrolytic rate of MA largely depends on concentration, on temperature, on pH-value and on additives which influence the reversible equilibrium. The time period in which 50% decrease by hydrolysis is 75 minutes with an initial concentration of 0.02 M of MA, but it is only 8 minutes with an initial concentration of 0.001 M. The addition of Ellman's reagent (5,5′-dithio-bis(2-nitrobenzoic acid)) causes the immediate oxidation of mesna to its disulfide and prevents the back reaction of mesna with 4-hydroxy-CP to MA [35]. Therefore, the half life time of MA is only 28 seconds. The addition of mesna reduces the hydrolysis, i.e. MA and its epimer are stabilized, and, for instance after addition of a 5-fold equivalent of mesna, the half

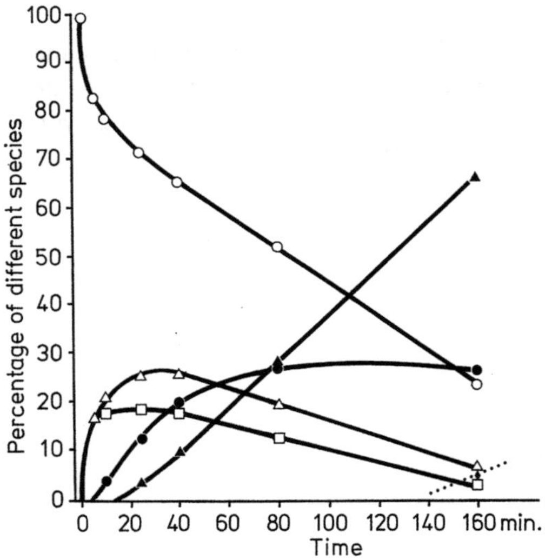

Fig. 7. Degradation of MA. Time course for MA plus *trans*-isomer (○), 4-hydroxy-CP (□), mesna (△), phosphoramide mustard (●), acrolein-mesna (1:1)-adduct (▲) and acrolein-mesna (1:2)-adduct (◆) in 0.07 M phosphate buffer, pH 7.0 at 37 °C, initially containing 0.02 M of MA. Procedures: photometric determination of 2-mercaptoethanesulfonate at pH 4.0 and 8.0 by 4,4′-dithiopyridine for MA, mesna and acrolein-mesna-(1:1)-adduct, semiquantitative thin layer chromatography on silicagel; detection: phorphorus-test, aldehyde-test and iodine-vapor for 4-hydroxy-CP and phosphoramide mustard

life time rises from 75 to 1,100 minutes. In the presence of another thiol, like 3-mercaptopropanesulfonate, the hydrolysis rate is also reduced in the same manner, but simultaneously with the epimerization the "thioethermoiety" is exchanged by transthioacetalization at the C-4 position of the oxazaphosphorine ring [35] according to the equation

$$\text{4-SR-CP} + \text{H-SR}' \rightleftharpoons \text{4-SR}'\text{-CP} + \text{H-SR}$$

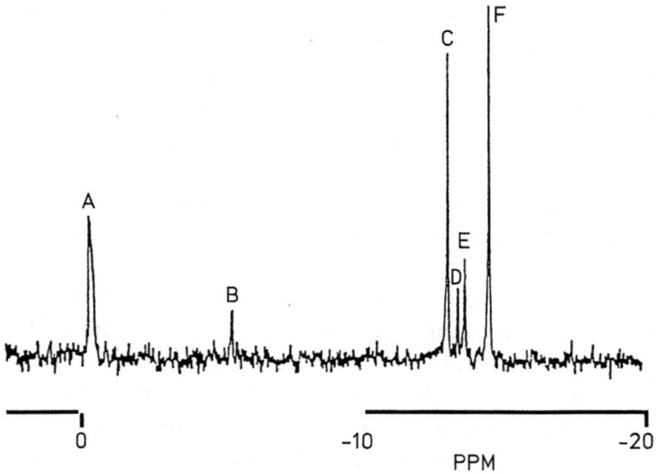

Fig. 8. Representative ^{31}P NMR spectrum of MA equilibrated in acetate buffer (100 mM, pH 5.55, 37 °C): A) triphenylphosphine oxide (coaxial reference in toluene-d_8), B) aldophosphamide and equivalents, C) *trans*-MA, D) *trans*-4-hydroxy-CP, E) *cis*-4-hydroxy-CP, F) MA

The activation mechanism of MA was elucidated by using ^{31}P-NMR-spectroscopy [41]. The hydrolysis products are well separated (Fig. 8). These experiments indicate that MA is activated by base-catalyzed elimination of the thio-moiety to yield imino-CP as a transient intermediate (Fig. 5). In vivo imino-CP can react with sulfhydryl-containing compounds to 4-thio-CP conjugates or with water to 4-hydroxy-CP which breaks down via aldophosphamide to phosphoramide mustard. The degradation of aldophosphamide is suppressed under more acidic conditions. Below pH 7 aldophosphamide reacts competitively with thiols to form hemithioacetal intermediates that subsequently cyclicize to the oxazaphosphorine ring system (Fig. 5) [41]. This capture of aldophosphamide by mesna may explain the bladder protective effect when CP and mesna are concomitantly given in animals [3] or patients.

3.4 The pH-Profile of Hydrolysis

The examination over a range from pH 2 to pH 9 allows the determination of that optimal pH-value giving the greatest stability. At about pH 4 hydrolysis, epimerization, transthioacetalization and the irreversible degradation to phosphoramide mustard are at a minimum (Fig. 9). Out of this range the stability decreases exponentially [42]. This behavior is utilized for the preparation of vials of MA by adjusting the solution at about pH 4 before lyophilization.

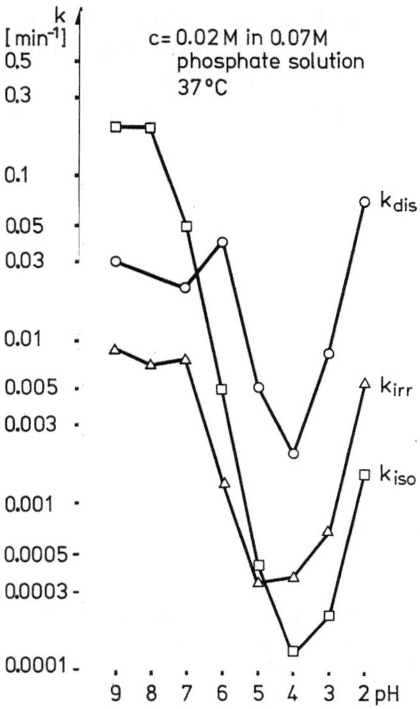

Fig. 9. Effect of pH on the rate constant for the disappearance of MA (c = 20 mM/l, in 0.07 M phosphate solution at 37 °C), rate constants: k dis (○) = reversible dissociation of MA to 4-hydroxy-CP and 2-mercaptoethanesulfonate, k iso (□) = isomerization of MA to the *trans*-isomer, k irr (△) = irreversible degradation of MA to phosphoramide mustard and acrolein

3.5 Stability of the Lyophilisate

MA-L-lysine as 100 mg lyophilisate in vials has sufficient storability at least at 4 °C. Under these conditions only 3% of the content decreased within two years. During storage at room temperature for two years approximately 10% of the drug was degraded. In contrast, 4-hydroxy-CP can only be stored for a few hours at 4 °C. Thus, MA is indeed a stabilized derivative.

A 10% hydrolysis of the reconstituted lyophilisate as a 0.5% aqueous MA solution occured at room temperature within four hours. Storage of the solution at +4 °C or —25 °C prolonged this time period to 24 hours or several months, respectively. The addition of a 5-fold equivalent of mesna at room temperature increased the time of 10% hydrolysis by up to eight days.

In conclusion, MA fulfils the following physico-chemical criteria for a suitable prodrug of 4-hydroxy-CP:

1) Sufficient stability in solid state;
2) good water solubility;
3) stability in solution at pH 4 before injection;
4) rapid liberation of 4-hydroxy-CP at pH 7 in vivo;
5) rapid release of the antidote mesna.

4 Pharmacokinetics

The schema of Fig. 10 is based on data of plasmakinetics and of urinary excretion after i.v. application of MA in rats. The urine was analyzed by a convenient thin-layer chromatographic method (TLC), which allows one to determine the main products directly and semiquantitatively (Fig. 10). Within the first 4 hours MA and

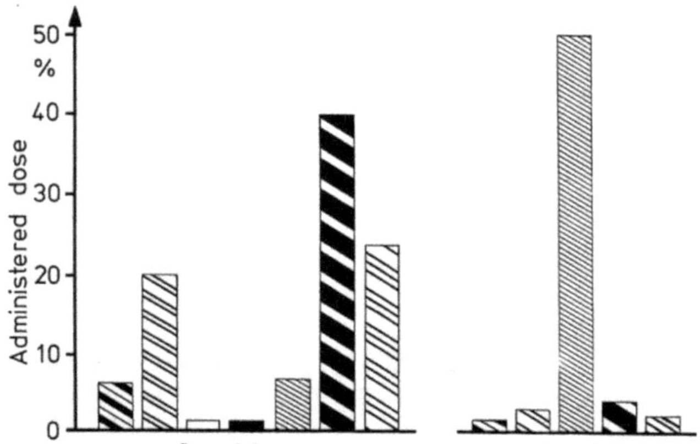

Fig. 10. Cumulative urinary excretion of MA plus trans-MA (▧), carboxyphosphamide (▧), keto-CP (□), phosphoramide mustard (■), phosphate (▧), mesna (▧) and dimesna (▧) in 4-hour-fractions after single i.v. administration of 500 μmol/kg of MA in the rat. Procedure: semiquantitative thin layer chromatography, detection: phosphorus-test and iodine-vapor

the epimer were found in a 1:1 ratio with about 6% of the administered dose, carboxyphosphamide with about 20% and roughly 7% phosphate. 4-keto-CP and phosphoramide mustard are minor metabolites with 1% each. Later on the phosphorus of the administered MA is nearly exclusively excreted as phosphate. Concerning alkylating activity detected by the NBP-test, about 30% of the whole dose were found in urine, i.e. about 70% were bound to biomolecules in vivo. The thio-substituent at C-4 of the oxazaphosphorine ring is predominantly eliminated as mesna with about 40% and dimesna with about 24%[43]. The high excess of urinary mesna in relation to the much lower amount of free 4-hydroxy-CP (far below 1%) prevents urotoxicity of MA. By photometric methods the amount of free thiols and reducible disulfides and of the masked acrolein from MA, 4-hydroxy-CP and isomers were unspecificly measured within half an hour or one hour fractions showing that the elimination is almost complete within two hours (Fig. 11)[44]. After four hours the cumulative excretions of MA plus isomer, mesna and dimesna are 6.5%, 43%, and 24%, respectively, in accordance with the data of the TLC-method.

The comparison of MA with CP shows that after i.v. administration similar amounts of 4-hydroxy-CP, respectively MA and its epimer are excreted (Fig. 11). This result is supported by the plasma levels (Fig. 12) showing beyond one hour nearly identical plasma concentrations. The initial half-life time of masked acrolein in rats is calculated at 33 minutes[44]. These elimination kinetics were identical for the cyclohexylamine salt and lysine salt of MA.

The direct analysis of MA in biological samples is hampered by the lack of a sensitive HPLC-procedure. A novel tool for direct identification and quantification of MA in urine or serum is ion-pair-extraction followed by positive or negative ion fast atom bombardement mass-spectrometry[45]. MA was found in urine samples of rats after intravenous administration as 1.5 + 3.6 + 1.4 + 0.4% of the injected dose during sampling periods for 0 to 1, 1 to 2, 2 to 4 and 4 to 6 hours, respectively.

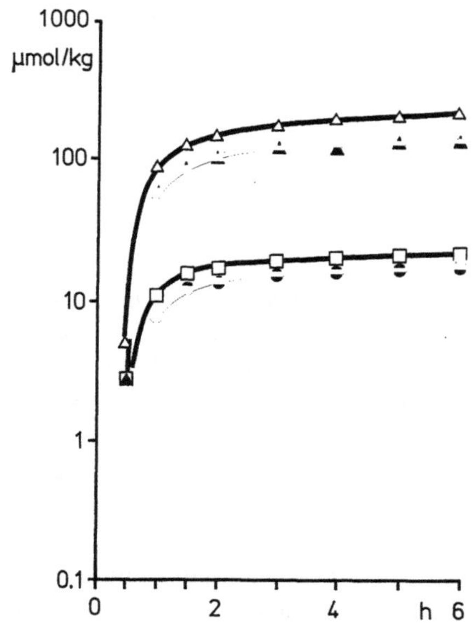

Fig. 11. Cumulative excretion of 4-hydroxy-CP (□) after single i.v. administration of 250 µmol/kg of CP and cumulative excretion of MA plus *trans*-epimer (●), mesna (▲) and mesna plus dimesna (△) after single i.v. administration of 250 µmol/kg of i.v. MA cyclohexylamine in the rat

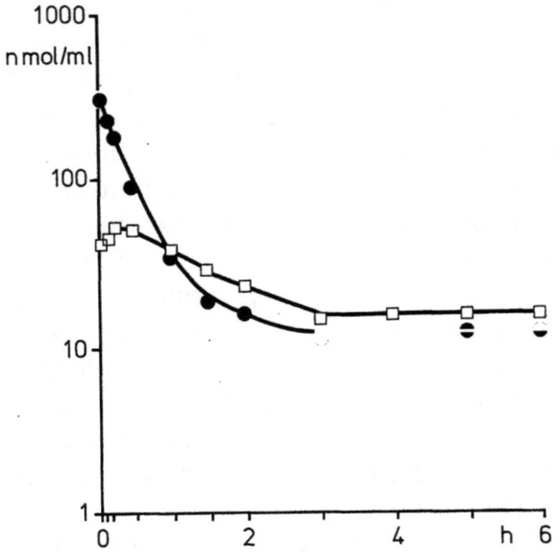

Fig. 12. Plasma concentrations (μmol/ml) of 4-hydroxy-CP respectively of masked acrolein in the rat after single i.v. administration of 250 $\mu mol^{-1}/kg^{-1}$ of CP monohydrate (□) or of MA cyclohexylamine (●)

With ^{31}P-NMR spectroscopy it was shown that the decomposition of MA at pH 7.4, 37 °C is much faster in plasma than in aqueous phosphate buffer with an enhanced appearance rate of phosphoramide mustard [41]. The behavior of MA was further studied by ^{31}P spectroscopy utilizing a human histiocytic lymphoma cell line (U937) [17], showing that MA did not enter the cell in contrast to 4-hydroxy-CP [46].

Due to variable infusion rates in human studies, only approximative pharmacokinetic parameters are available. At MA doses below 2 mg/m^2 concentration profiles of masked acrolein (MA/4-hydroxy-CP) are compatible with a monoexponential decay with T1/2 about 8 minutes while at higher doses a biexponential model results with a similar initial phase and a longer lasting β-phase with T1/2 about 2 hours [47].

The decline of mesna plasma levels deriving from a concomitant MA and mesna infusion showed a half-life time in the range of 36 to 72 minutes while the excretion of dimesna was somewhat slower with half-life times ranging from 1.1 to 3.2 hours [47].

5 Mode of Action

The mode of action of MA is closely related to its metabolism as outlined above. There is multiple evidence that the major lesion induced by a bifunctional alkylating agent, responsible for cytotoxicity, is DNA interstrand crosslinking [48]. MA has been shown to induce ISC in rat bone marrow, murine L1210 and human leukemic and lymphoid cells [49]. Among patients a heterogeneity in sensitivity to MA was observed. However, sequential analyses of patients at different stages of their disease revealed a parallel decline of ISC formation and chemosensitivity of their malignant disease. Differences in DNA repair capacity are discussed [50].

Inhibition of histone acetylation as demonstrated for Ehrlich Ascites Tumor cells may be another mode of action of MA [51]. Furthermore, these investigations have shown that oxazaphosphorines can also have an alkylating action on the cell membrane; some special immunological phenomena, e.g. shedding of membrane receptors or structures, are possibly triggered by direct, non-alkylating interactions with the cell membrane [52,53].

6 Experimental Characterization of MA

6.1 Pharmacological Activity

MA is a cytotoxic compound active in vitro and in vivo, the presumed mode of action being identical to that of CP. MA is highly effective in vitro against established rodent tumor cell lines at concentrations of 1 to 10 μg/ml in a two-hour incubation assay. This is in sharp contrast to CP, which is absolutely ineffective in vitro. The in vivo activity of MA against a broad spectrum of rat and mouse tumors is well documented with an optimal therapeutic dose range between 50 mg/kg and 200 mg/kg in mice and rats [1,54–56]. Cross-resistance against CP appears to be incomplete when MA was tested in a CP-resistant line of the P 388 mouse leukemia [57]. The antitumor activity of MA covers the same broad spectrum of experimental tumors in mice and rats as CP and the optimal doses of CP and MA are similar on a mg basis. On a molar basis, however, MA is almost twice as active as CP. Results in the Lewis lung tumor [58] support the superiority of dosis fractionation, whereas in the L 1210 leukemia no dependence on the administration scheme was observed. The therapeutic index, determined as the ratio of the LD 10 and of the dose necessary for a 40% prolongation of the survival time in the P 388 leukemia was higher by a factor of about 3 for MA than for CP. Similar results were obtained in a rat leukemia model [59]. MA caused a retardation of growth in 6 transplants of 10 human ovarian carcinomas in the nude mouse and tumor regression in the other 4 cases [60]. In other human xenografts in nude mice MA proved activity against a malignant melanoma, but no activity against a stomach carcinoma [58]. In addition, activity against testicular cancer xenografts has been reported [61].

6.2 Experimental Toxicology of MA

The LD 50 for MA in the mouse is between 500 mg/kg and 625 mg/kg on i.v. administration and 2.310 mg/kg orally [44]. A fractionated dosage scheme [58] was less toxic, compared with a single i.p. dose. For the rat the LD 50 values (i.v.) are in the range of 250 to 310 mg/kg and orally at 1.000 to 1.250 mg/kg. In the rat, a second dose of MA given after the disappearance of all toxic symptoms from the first dose, shows a qualitatively and quantitatively identical toxicity profile to that after the first injection. The LD 50 values in rats and mice with single i.v. or oral administrations correspond well with data for CP, but the cumulative toxicity (70%) in rats following fractionated i.v. injections on four consecutive days is less pronounced than that of CP (>95%) [44].

In contrast to the enhanced antitumor activity of MA, the adverse side effects of CP such as leukocyte depression, imunosuppression and urotoxicity are considerably lower with MA. Especially, urotoxic symptoms were not observed. This might be caused by the excess of mesna versus 4-hydroxy-CP in the bladder, the spontaneous re-formation of MA and the prevention of degradation to acrolein respectively (Chapt. 3.3 and 4).

The primary sign of toxicity of MA in mouse and rat is myelosuppression. Dose-dependent leukocyte nadirs occur in the rat on the 3rd day after i.v. MA; recovery is complete by day 10 [44]. The time courses for the action of i.p. doses of MA and CP on haemopoietic stem cells were compared in the mouse [62]. On an equitoxic dosage (LD 10) the suppression of CFU-C, CFU-S, CFU-E, and BFU-E, was more pronounced after CP than after MA; recovery after MA was faster and was already complete after 8 days.

Significant immunosuppression was observed only at very high doses, whereas lower doses of MA had no suppressive effects on production of humoral antibodies, but produced slight, but significant, increases of titres during 3 weeks (Table 1) [55].

DNA cross-linking and single strand breaking was found in the bone marrow of MA treated rats [42]. Tissue irritation may occur after intradermal injections of concentrations above 2 mg/ml [55].

Table 1. Hemolytic antibody titres (mean log 2) in rats over 21 days after SRBC immunization and concomitant treatment with MA or CP

Dose mg/kg	Mafosfamide × log 2	Cyclophosphamide × log 2
—	7.4	7.4
10.0	7.9	7.0
21.5	8.4	3.9
46.4	7.0	1.6
100.0	1.5	—

A special feature of MA was first observed clinically as immediate and reversible pain along the injected vain, but without objective irritation. An experimental rat model for the study of venous pain induced by 4-hydroxy-CP derivatives was developed and validated. Using various metabolites and chemical variants of 4-hydroxy-CP it was found that pain induction was independent of the compounds alkylating activity but possibly related to the spontaneous generation of minute amounts of acrolein from the 4-hydroxy-CP molecule. Accordingly, the pain could be prevented by the addition of thiol compounds such as mesna or *N*-acetyl-cysteine [63].

7 Clinical Investigations with MA

7.1 *Systemic Administration*

Phase I studies were performed with single intravenous doses [47,64,65] with weekly intravenous doses [66] in advanced solid tumors as well as with 5-day intravenous doses every four weeks. Subjective pain along the infused vein, however without clinical signs of phlebitis, was observed with doses above 300 mg/m^2. Additionally, unpleasant sensations occurred in the mucous membranes of nose and throat. These symptoms were dose-limiting whereas no hematological or other toxicity was clinically significant up to this dose. Concurrent administration of MA and mesna stopped the sensation of pain in practically all patients [67]. Solutions with concentrations of MA (6 mg/ml) and mesna (9 mg/ml) were tolerated at an i.v. infusion rate of 4 to 5 ml/minute [68] corresponding to the infusion of 25 to 30 mg/minute (equivalent to 1,500 to 1,800 mg MA/hour). With this modified procedure the maximum tolerated total dose of MA could be increased to 3,000 mg/m^2/course [47]. Hematological toxicity was found at a MA regimen of 5q 700 mg/m^2/week [66] or at single 3,000 mg/m^2/course [47,64]. Nausea and vomiting were observed as non-hematological toxicity. The maximum tolerated dose of MA on fractionated use and without addition of mesna is under investigation in a further phase I study.

7.2 *MA in Local/Regional Therapy*

Experimentally, this activity of MA was shown with intracavitary administration in the belly bath model on rats [69] and with intra-arterial infusion in the Novikoff hepatoma of the rat [70,71]. Compared with an i.v. infusion the i.a. infusion caused a greater tumor regression with an increased proportion of complete remissions. In pilot clinical investigations MA was employed intrapleurally for malignant pleural effusion and intraperitoneally for ascites as well as for peritoneal carcinosis [72]. Local pain reactions or burning occurred, however the symptoms could be controlled by slowing the rate of infusion.

7.3 *MA for Predictive Sensitivity Testing*

An effective pre-therapeutic drug sensitivity testing of individual tumors could save an individual cancer patient from a physically and mentally exhausting but ineffective therapy and has therefore recently become of increasing significance for the planning and evaluation of tumor chemotherapy in terms of the "quality of life". In the case of CP and ifosfamide, which are essential constituents of many chemotherapeutic regimes, it is not possible to individually test these drugs pretherapeutically. Here MA can be used as an indicator for sensitivity or resistance to CP and to structurally-related alkylating agents. MA was studied in the colony assay of Hamburger and Salmon [73] using fresh explants of human tumors [74,75]. Sensitivity was found for ovarian carcinomas and other tumors of various origins, but no correlation with in vivo therapy results was undertaken. In a comparison of MA with 4-hydroperoxy-CP in mammary carcinoma explant material it was shown that a correlation exists between the results with the two CP derivatives: concordant resistance in 2 out of 23 cases, concordant sensitivity in 12 out of 23 cases, and disordant results in 9 out of 23 cases but without discernible preference for one of the two derivatioves [76]. In a biochemical short-term assay in ovarian carcinoma samples both compounds proved to be equally effective, with concordant sensitivity in 12 assay pairs [76, 77].

7.4 *Bone Marrow Purging in Autologous Bone Marrow Transplantation (ABMT)*

4-hydroperoxy-CP has previously been shown to specifically eliminate leukemic stem cells from an excess of normal bone marrow without concomitant loss of the capability of pluripotent stem cells to regenerate normal hemopoiesis [78]. The stem-cell sparing effect of 4-hydroxyperoxy-CP [79] is attributed to a higher content of aldehyde dehydrogenase in these cells [80].

The stem-cell sparing effect or, additionally, an elimination of tumor cells from bone marrow cell suspensions, was also shown for MA by several groups [62, 81–84].

A combination of MA with other agents such as another cytostatic [85], monoclonal antibodies/immunotoxins [86–90] or photosensitizers [91] leads to a further augmentation of the purging effect. These approaches however, apart from MA/monoclonal antibodies/complement [86] have not yet been employed clinically.

ABMTs with MA purging have been performed by a number of transplantation teams in Europe, including national [92–93] or international study groups (European Bone Marrow Transplantation Group) [94]. Acute leukaemia predominates as indication [92, 94–96]; lymphoma and neuroblastoma were further indications. In total, about 400 patients have received MA-treated autologous bone marrow transplants. Spontaneous and only slightly delayed recovery is seen even in the absence of detectable colony forming units of granulocyte/marcophage stem cells in the transplant. Thrombocyte recovery appears to be more sensitive to MA-purging than leucocyte recovery; other purging related toxicities have not been reported.

Favorable clinical results have been reported by teams with larger series of purged transplants in partients with acute myologenous leukemia [94, 95]. Summarizing a recent analysis of the European bone marrow transplantation group [94, 97] and of an Italian cooperative group [93], results that patients with acute lymphocytic leukemia show a trend in favour of MA purging as compared with non-purged ABMTs. A final assessment of MA purging and its contribution to the general benefit of ABMT must await results from prospectively randomized studies. The available results, however, have impressively proven the feasibility of this pharmacological purging approach.

8 Immunmodulation with Oxazaphosphorines: Future Aspects for MA

Immunosuppression is a well-documented side-effect of CP [98], which is therapeutically utilized in certain clinical situations. In some experimental studies, however, an apparently paradoxical effect was observed: an increase in the immune response by low CP doses [99-101]. Low CP doses also increased delayed-type cutaneous sensitivity reactions [102].

The first experimental model in which the immunological efficacy of low doses of CP on tumor growth was detected and quantified is the transplantable monoytic rat leukemia L5222 [103]. This tumor is curable by low-dose CP or MA [104]. The dose-effect relationships for CP and for MA, are shown in Fig. 13. A bell-shaped dose-effect relationship results for both substances with the optimal therapeutic effect in a low dose range. An essential difference between the two substances is, however, the considerably greater dose range of MA in which a curative effect can be obtained.

The analysis of surface markers in stimulated lymphocyte cultures indicated that the occurrence of T suppressor populations was inhibited by the presence of low concentrations of 4-hydroperoxy-CP [105]. A similar phenomenon was also detectable in patients treated with differing CP doses [106]. This is the basis for the hypothesis that low CP doses in vivo lead to a selective inhibition of T suppressor cell activity, causing the observed immunostimulating action [107]. Suppressor T cell subsets moreover vary in their sensitivity to CP, and immature precursor cells seem to be more sensitive than mature T cells [102]. The therapeutic efficacy of low immunomodulatory CP doses can be shown in several experimental tumor models. The MOPC-315-plasmocytoma of the Balb/c mouse was especially suited for studying the immunopharmacology of CP. In this transplantation tumor it was shown for the first time that the reproducable antitumor activity of CP is directly related to an effect on T cells [108]. The curative effect of low CP doses could be completely abolished by simultaneous administration of mouse antithymocyte serum, which causes the elimination of all T cells [109].

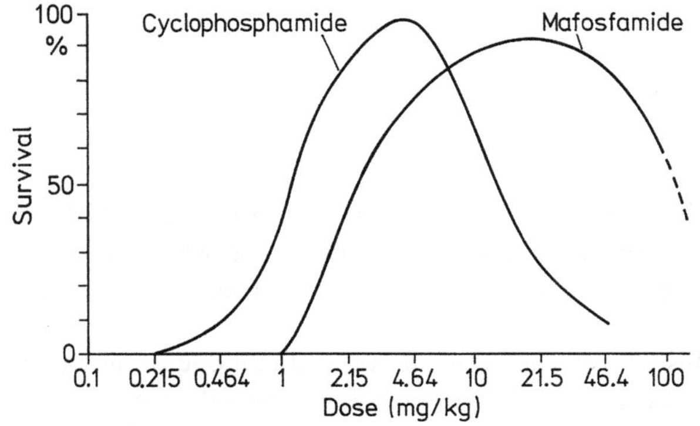

Fig. 13. Curative dose-efficiency curves in the L 5222 leukemia of the rat

Further investigations on the rat leukemia L5222 have proven that cures by low oxazaphosphorine doses depend on activation of host mechanisms: animals which were cured by the maximum tolerated dose of CP (above 100 mg/kg) or MA (above 200 mg/kg), are killed by the leukemia after renewed tumor implantation. In contrast, a complete resistance towards renewed tumor transplantation exists in those rats which were cured at the low dose range [110]. The experimental data now available show that regulatory T cells are more sensitive towards CP and MA than are T effector or T helper cells. The molecular bases of this selective action are as yet largely unknown. However, they cannot depend exclusively on the capability of oxazaphosphorines to form DNA crosslinks. Most other cytostatics which interfere with DNA do not possess this immunomodulatory effect. A monofunctional CP derivative, dechloroethyl-CP, on the other hand, also acts in the MOPC-315 model in a positive immunomodulatory manner [111], which again indicates that the T cell-specific properties of the molecule are probably unrelated to DNA crosslinking.

There are many clinical findings indicating that an immunomodulation could also be therapeutically useful, although up to now the reproducability and constancy of the results of the immunological tests during CP therapy have been generally poor. This may derive from, among other things, the relatively low immunotherapeutic range of the drug, which again should be related to its metabolism and pharmacodynamics. However, MA has the necessary greater immunotherapeutic range and it does not require a biotransformation, which means that the plasma level is predictable. In comparison with CP, the immunosuppressive properties of high MA doses are also considerably less marked and the lower bone marrow toxicity of this compound makes it an important candidate for clinical trials as an immunopharmaceutical. Clinical phase I studies aimed at exploring the immunomodulating effect of low-dose MA showed an improved immune status of cancer patients following i.v. administration of 25 mg/m^2 MA. This was manifested as a stimulation of delayed type immune reactions (Multitest Mérieux) in patients with depressed pretreatment levels. A pronounced elevation of lymphocyte surface markers for T lymphocytes, natural killer cells and for macrophages, was observed after MA administration, with a peak between day 5 and day 9 [112, 113]. No toxic effects of MA have been encountered at low doses of MA, and recently phase II studies have been initiated.

9 Conclusions

MA-L-lysine was developed in the series of oxazaphosphorine cytostatics, e.g. CP and ifosfamide, as a prodrug of the systemically effective 4-hydroxy-CP which is hydrolytically released as well as of the locally active mesna for the protection of kidney and bladder.

The chemical synthesis of this CP-mesna conjugate was difficult due to the initial isomerization and to the hydrolytically unstable, not crystallizable salt-like structure of MA L-lysine. These problems were solved by a stereoselective synthesis of MA cyclohexylamine salt and by a time-, temperature- and pH-optimized cation exchange with subsequent lyophilization. The storability of this lyophilisate is sufficient at 4 °C. Therefore, MA can be characterized as a "stabilized" derivative of 4-hydroxy-CP.

The stereochemistry of MA was determined by polarimetry, X-ray analysis and NMR spectroscopy, proving that MA is a racemic *cis*-isomer with a diaxial pattern of the phosphoryl oxygen and the thio-substituent at C-4. The rate constants of decomposition in aqueous solution show that the isomerization to *trans*-MA, the hydrolysis to 4-hydroxy-CP/2-mercaptoethanesulfonate and the irreversible degradation to phosphoramide mustard/acrolein is at a minimum at pH 4. Imino-CP was confirmed as a transient intermediate for the rapid basecatalyzed reaction of MA with thiols to 4-thio-CP-derivatives or with water to 4-hydroxy-CP.

Pharmacokinetic studies show after i.v. application of MA in rats the serum half-life time of MA/4-hydroxy-CP is about 33 minutes. The main metabolites in urine are carboxyphosphamide, mesna, dimesna and MA plus *trans*-isomer. It was suggested that renal-eliminated 4-hydroxy-CP and mesna react via aldophosphamide hemithioacetal to the isomeric mixture of MA/*trans*-MA in urine. This reaction prevents the liberation of acrolein and therefore no urotoxic symptoms were observed after administration of MA.

MA opens up new possibilities in clinical therapy and in pretherapeutic tests, for examination in the clonogenic stem cell assay, for in vitro purging for ABMT, and for the regional therapy of tumors. A completely new type of therapy is emerging, specifically for low-dose MA, as immunomodulator, possibly in combination with further substances from the biological response-modifier group.

10 Acknowledgements

The development and characterization of MA was partly supported by the Bundesministerium für Forschung und Technologie.

The technical assistance of P. Bäumer, M. Elliger, E. Kampherm, M. Meyer zur Heyde, B. Piening, H. Skupzig and K. Uffmann is appreciated. Thanks are given to K. Linnenkamp and A. Obermeier for skillful preparation of the manuscript.

11 Abbreviations

MA = Mafosfamide = (±)-2-[[2-[Bis(2-chloroethyl)amino]-tetrahydro-2H-1,3,2-oxazaphosphorine-4-yl]thio]ethanesulfonic acid-P-*cis*-oxide
CP = Cyclophosphamide
ABMT = Autologous bone marrow transplantation
INN = International non-proprietory name

12 References

1. Brock N (1983) Oxazaphosphorine cytostatics: structure-activity relationships, selectivity and metabolism, regional detoxification. In: Reinhoudt DN, Connors TA, Pinedo HM, van de Poll KW (eds) Structure-activity relationships of anti-tumor agents. Developments in pharmacology 3; Martinus Nijhoff, The Hague, p 239
2. Brock N., Hohorst H-J (1963) Arzneim.-Forsch./Drugs Res. 13: 1021
3. Hohorst H-J, Ziemann A., Brock N (1971) Arzneim.-Forsch./Drug Res. 21: 1254
4. Brock N, Hohorst H-J (1977) Z. Krebsforsch. 88: 185
5. Brock N, Stekar J, Pohl J, Niemeyer U, Scheffler G (1979) Arzneim.-Forsch./Drug Res. 29: 659
6. Brock N (1980) New experience with the oxazaphosphorines with special reference to the uroprotector Uromitexan. In: Burkert H, Nagel GA (eds) Contributions to Oncology 5; S. Karger, Basel, p 1
7. Borch RF, Hoye TR, Swanson TA (1984) J. Med. Chem. 27: 490
8. Voelcker G., Draeger U, Peter G, Hohorst H-J (1974) Arzneim.-Forsch./Drug Res. 24: 1172
9. Friedman OM, Myles A, Colvin M (1979) Cyclophosphamide and related phosphoramide mustards: current status and future prospects. In: Rosowsky A (ed) Advances in Cancer Chemother., Dekker, New York, p 143
10. Friedman OM (1967) Cancer Chemother. Rep 51: 347
11. Engle TW, Zon G, Egan W (1979) J. Med. Chem. 22: 897
12. Grunicke H, Doppler W, Hofmann J, Lindner H, Maly K, Oberhuber H, Ringsdorf H, Roberts JJ (1986) Advances in Enzyme Regul. 24: 247
13. Wrabetz E, Peter G, Hohorst H-J (1980) J. Cancer Res. Clin. Oncol. 98: 119
14. Watson E, Phoebe D, Chan KK (1985) J. Pharm. Sci. 74: 1283
15. Draeger U, Hohorst H-J (1976) Cancer Treat. Rep. 60: 423
16. Lenssen U, Hohorst H-J (1979) J. Cancer Res. Clin. Oncol. 93: 161
17. Boyd VL, Robbins JD, Egan W, Ludeman SM (1986) J. Med. Chem. 29: 1206
18. Zon G, Ludeman SM, Sweet EM, Egan W, Phillips LR (1982) J. Pharm. Sci. 71: 443
19. Sladek NE (1973) Cancer Res. 33: 651
20. Zon G, Ludeman SM, Brandt JA, Boyd VL, Özkan G, Egan W (1984) J. Med. Chem. 27: 466
21. Borch RF, Gutman KM (1984) J. Med. Chem. 27: 485
22. Draeger U, Peter G, Hohorst H-J (1976) Cancer Treat. Rep. 60: 355
23. Low JE, Borch RF, Sladek NE (1983) Cancer Res. 43: 4815
24. Bielicki L., Voelcker G, Hohorst H-J (1984) J. Cancer Res. Clin. Oncol. 107: 195
25. Connors TA, Cox PJ, Farmer PB, Foster AB, Jarman M (1974) Biochem. Pharmacol. 23: 115
26. Brock N, Stekar J, Pohl J, Scheef W (1979) Naturwissenschaften 66: 60
27. Scheef W, Klein HO, Brock N, Burkert H, Günther U, Hoefer-Janker H, Mitrenga D, Schnitker J, Voigtmann R (1979) Cancer Treat. Rep. 63: 501
28. Brock N, Hilgard P, Pohl J, Ormstad K, Orrenius S (1984) J. Cancer Res. Clin. Oncol. 108: 87
29. Cox PJ (1979) Eur. J. Cancer 15: 1071
30. Brock N (1986) J. Cancer Res. Clin. Oncol. 111: 1
31. Peter G, Hohorst H-J (1979) Cancer Chemother. Pharmacol. 3: 181
32. Ramonas LM, Erickson LC, Ringsdorf H, Zaharko DS (1980) Cancer Res. 40: 3704
33. Takamizawa A, Matsumoto S, Iwata T, Kentagiri K, Tochino Y, Yamaguchi K (1973) J. Amer. Chem. Soc. 95: 985
34. Peter G, Wagner T, Hohorst H-J (1976) Cancer Treat. Rep. 60: 429
35. Niemeyer U, Engel J, Scheffler G, Molge K, Sauerbier D, Weigert W (1984) Invest. New Drugs 2: 133
36. Niemeyer U, Engel J, Scheffler G, Pohl J, Hägele G, Krebs B (1987) Phosphorus and Sulphur 30: 585
37. White DW, Gibbs DE, Verkade JG (1979) J. Amer. Chem. Soc. 101: 1937
38. Setzer WN, Sopchick AE, Bentrude WG (1985) J. Amer. Chem. Soc. 107: 2083
39. Domeyer BE, Sladek NE (1980) Biochem. Pharmacol. 29: 2903

40. Molge K (1983) Characterization of the hydrolysis of mafosfamide. ASTA Pharma AG, Research Report no. 53
41. Kwon C-H, Borch RF, Engel J, Niemeyer U (1987) J. Med. Chem. 30: 395
42. Niemeyer U, Engel J, Meyer zur Heyde M, Molge K, Wolf-Heuss E (1987) International Symposium on pharmaceutical and biomedical analysis, Barcelona, p 244
43. Niemeyer U (1978) TLC-determination of mafosfamide and metabolites in urine samples. ASTA Pharma AG
44. Pohl J, Hilgard P, Jahn W, Zechel H-J (1984) Invest. New Drugs 2: 201
45. Manz I, Dietrich I, Przybylski M, Niemeyer U, Pohl J, Hilgard P, Brock N (1985) Biomed. Mass Spectrometry 12: 545
46. Ludeman SM (1986) Personal communication
47. Burton LC, Harper PG, James CA, Rogers HJ, Liang R (1985) The clinical trial and pharmacokinetics of 4-hydroxycyclophosphamide and mesna derived from mafosfamide (ASTA Z 7557) and its comparison with other oxazaphosphorines in man. Report to Cancer Research Campaign
48. Osieka R, Pannenbächer R, Schmidt CG (1984) Invest. New Drugs 2: 161
49. Berger MR, Bedford P, Zeller WJ, Kaufmann M (1984) Invest. New Drugs 2: 181
50. Huang AT, Kremer WB, Laszlo J, Setlow RB (1972) Nature New Biol. 240: 114
51. Grunicke H, Gruenewald K, Helliger W, Scheidl F, Wolf-Schreiner E, Puschendorf B (1983) Adv. in Enzyme Regul. 21: 21
52. Askenase PW, Hayden BJ, Gershon RK (1975) J. Exp. Med. 141: 697
53. L'Age-Stehr J, Diamantstein T (1978) Nature 271: 663
54. Brock N, Niemeyer U, Pohl J, Scheffler G (1981) Third NCI-EORTC symposium on new drugs in cancer therapy, No. 47, Brussels
55. Pohl J (1984) Mafosfamide — Summary for Investigators. Asta Pharma AG
56. Atassi G, Hilgard P, Pohl J (1984) Invest. New Drugs 2: 169
57. Brock N, Hilgard P, Peukert M, Pohl J, Sindermann H (1989) Cancer Invest.
58. Klein HO, Wickramanayake PD, Christian E, Coerper C (1984) Invest. New Drugs 2: 191
59. Zeller WJ, Berger MR, Matys R, Schumacher J (1984) Invest. New Drugs 2: 175
60. Kleine W, Düser B, Runge M, Grezes IR, Kröger H, Klosa J (1984) J. Cancer Res. Clin. Oncol. 107: 80
61. Osieka R, Glatte P, Bamberg M, Scherer E, Schmidt CG (1984) Proc. Am. Ass. Canc. Res. 25: 330
62. Nowrousian MR, Schmidt CG (1984) Invest. New Drugs 2: 207
63. Hilgard P, Pohl J (1986) Invest. New Drugs 4: 373
64. Bruntsch U, Groos G, Hiller TA, Wandt H, Tigges FJ, Gallmeier WM (1985) Invest. New Drugs 3: 297
65. Bruntsch U, Groos U, Hiller TA, Wandt H, Tigges FJ, Gallmeier WM (1984) Cancer Immunol. Immunother. 18: 11
66. Abele R, Aapro MS, Haeflinger JM, Alberto P (1986) Cancer Chemother. Pharmacol. 16: 182
67. Peukert M (1984) Preliminary clinical evaluation of mafosfamide. ASTA Pharma AG
68. Peukert M (1985) Phase-I studies with mafosfamide. ASTA Pharma AG
69. Voelcker G, Laber P, Rockinger H, Wientzek C, Hohorst HJ (1984) Invest. New Drugs 2: 253
70. Friedl P, Berger MR, Rozycki-Gerlach W, Schlag P, Schmähl D (1988) J. Cancer Clin. Res. Oncol. 114, Suppl. S 187
71. Bartkowski R, Berger MR, Aguiar JLA, Henne TH, Dörsam J, Geelhaar GH, Schlag P, Herfarth C (1986) J. Cancer Res. Clin. Oncol. 111: 42
72. Wagner T, Walter E, Mittendorf F, Voelcker G (1984) J. Cancer Res. Clin. Oncol. 107: 46
73. Hamburger AW, Salmon SE (1977) Science 197: 461
74. Alberts DC, Einspahr JG, Struck R, Bignami G, Young L, Surwit EA, Salmon SE (1984) Invest. New Drugs 2: 141
75. Einspahr J, Alberts DS, Struck R, Salmon SE (1984) Proc. Am. Soc. Clin. Oncol. 25: 38
76. Kaufmann M (1986) Personal communication
77. Buse KD, Kaufmann M, Kubli F (1985) Der biochemische Kurzzeit-Inkubations-Assay zur In-vitro Chemoresistenztestung und als Prognosefaktor bei menschlichen Tumoren. In: Jonat W, Kaufmann M, Kubli F et al. (eds) Aktuelle Onkologie 16, Zuckerschwerdt, München

78. Sharkis SJ, Santos GW, Colvin M (1980) Blood 55: 521
79. Körbling M, Hess AD, Tutschka PJ, Kaizer H, Colvin M, Santos GW (1982) Brit. J. Hematol. 52: 89
80. Kohn FR, Sladek NE (1984) Proc. Am. Ass. Canc. Res. 25: 289
81. Aglietta M, Stacchini A, Sanavio F, Piacibello W, Gavosto F (1987) Bone Marrow Transpl. 2: 56
82. Douay L, Gorin NC, Laporte J-P, Lopez M, Najman A, Duhamel G (1984) Invest. New Drugs 2: 187
83. Hagenbeek A, Martens ACM (1984) Invest. New Drugs 2: 237
84. Hervé P, Cahn JY, Plouvier E, Flesch M, Tamayo E, Leconte des Floris R, Peters A (1984) Invest. New Drugs 2: 245
85. Hervé P, Tamayo E, Cahn JY, Plouvier E, Flesch M, Peters A (1986) Attempts to eliminate residual acute myeloid leukemia from autologous bone marrow grafts through in vitro chemotherapy — a review. In: Hagenbeck A, Loewenberg B (eds) Minimal residual disease in leukemia, Martinus Nijhoff, Dordrecht, p 248
86. Hervé P, Plouvier E, Amsallen D, Racadot E, Noir A (1987) Bone Marrow Transpl. 2: 53
87. LeBien TW, Anderson JM, Vallera A, Uckun FM (1986) Leukemia Res. 10: 139
88. Uckun FM, Ramakrishnan S, Houston LL (1985) Blood 50: 19
89. Uckun FM, Ramakrishnan S, Haag D, Houston LL (1985) Leukemia Res. 9: 83
90. Uckun FM, Stong RC, Youle RJ, Vallera DA (1985) J. Immunol. 134: 3504
91. Porcellini A, Marchetti Rossi MT, Centis F, Manna A, Talevi N, Agostini A, Pazzaglia C (1987) Bone Marrow Transpl. 2: 52
92. Flesch M, Hervé P (1985) Blood Transfus. Immunohaemat. 28: 489
93. Rizzoli V, Mangoni L, Carella AM, Torlontano G, Aglietta M, Mozana R, Porcellini A, Coleselli P (1987) Exp. Hemat. Today (in press)
94. Gorin NC, Aegerter P (1987) Bone Marrow Transpl. 2: 320
95. Körbling M, Ho A, Dörken B, Haas R, Pezzutto A, Martin H, Deboben A, Reichardt P, Tischbirek K, Mende U, Fehrentz D, Hunstein W (1987) Bone Marrow Transpl. 2: 62
96. Laporte JP, Gorin NC, Douay L, Lopez M, Bauters F, Jouet JP, Walter MP, Lemonnier MP, Salmon C, Najman A (1987) Bone Marrow Transpl. 2: 61
97. Gorin NC, Hervé P, Aegerter P, Goldstone A, Linch D, Maraninchi D, Burnett A, Helbig W (1986) Brit. J. Haematol. 64: 385
98. Potel J, Brock N (1971) Arzneim.-Forsch./Drug Res. 21: 1250
99. Scheper RJ, Vos A, de Groot J, Boerrigter GH (1984) Invest. New Drugs 2: 221
100. Mitsuoka A, Baba M, Morikawa S (1976) Nature 262: 77
101. Turk JL, Parker D: The effect of drugs on immunological control mechanisms. In: Turk JL, Parker D (eds) Drugs and immune responsiveness, Macmillan, London, p 73
102. Ozer H, Cowens JW, Nussbaum A, Mihich E, Colvin M (1981) Fed. Proc. 40: 1075
103. Brock N (1978) Pharmakologische Grundlagen der Krebs-Chemotherapie. In: Georgii A (ed) Verhandlungen der Deutschen Krebsgesellschaft. Band 1, 13. Deutscher Krebskongreß Hamburg 1976. Gustav Fischer, Stuttgart, p 15
104. Hilgard P, Pohl J, Stekar J, Voegeli R (1985) Cancer Treatm. Rev. 12, 155
105. Cowens JW, Ozer H, Ehrke J, Colvin M, Mihich E (1981) Fed. Proc. 40: 1086
106. Berd D, Mastrangelo MJ, Engstrom PF, Paul A, Maguire H (1982) Cancer Res. 42: 4862
107. Duclos H, Galanaud P, Devinsky O, Maillot MC, Dormont J (1977) Eur. J. Immunol. 7: 679
108. Hengst JCD, Mokyr MB, Dray S (1980) Cancer Res. 40: 2135
109. Mokyr MB, Hengst JC, Dray S (1982) Cancer Res. 42: 974
110. Voegeli R, Pohl J, Reissmann T, Stekar J, Hilgard P (1986) Therapeutic efficacy of oxazaphosphorines by immunomodulation. In: Chandra P (ed) New Experimental modalities in the control of neoplasia. Plenum, p 243
111. Mokyr MB, Brundrett RB, Colvin M, Dray S (1986) Cancer Res. 46: 3313
112. Klein HO, Kreysch H-G, Coerper C, Voigt P, Ruff I (1987) Meth. and Find. Exptl. Clin. Pharmacol. 9: 627
113. Klein HO, Kreysch H-G, Coerper C, Voigt P (1988) J. Cancer Res. Clin. Oncol. 114, Suppl. S 126

Enzymatic DNA Methylation[1]

Stefan Grünwald and Gerd P. Pfeifer

Zentrum der Biologischen Chemie, Johann Wolfgang Goethe-Universität Frankfurt am Main, Theodor-Stern-Kai 7; D-6000 Frankfurt am Main, FRG.

The genome of mammals contains in addition to the four major DNA bases small amounts of 5-methylcytosine. The functional role of this DNA base has been the subject of numerous investigations especially during the last decade. This review is an attempt to summarize the present knowledge about synthesis, distribution and possible functions of 5-methylcytosine especially in eukaryotic genomes.

---

[1] Dedicated to the memory of Prof. Dušan Drahovsky

1 Introduction

In addition to the DNA bases adenine, guanine, cytosine and thymine, which are incorporated into DNA during replication, the DNA of mammalian cells contains low amounts of the base 5-methylcytosine, which arises after DNA synthesis by enzymatic methyl group transfers to certain cytosines. This minor base was first described by Hotchkiss [1] and Wyatt [2] as a rare component of DNA. Its possible role in regulation of gene expression was first suggested by Srinivasan & Borek [3]. This concept was elaborated by Riggs [4], and Holliday and Pugh [5]. After the development of recombinant DNA techniques a considerable progress was made in understanding the function of DNA methylation in mammalian cells. The matter has been discussed in several reviews [6-12] and books [13-15] in recent years. As a consequence of increasing research efforts in this field, new insights are being continuously provided. However, one has to admit that there are still more questions than answers.

2 DNA Methylation in Bacteria

DNA methylation is the most common form of DNA modification in prokaryotes. It has been extensively studied in a variety of prokaryotic organisms [16-18]. In contrast to eukaryotes, where 5-methylcytosine is the dominant minor base, at least three minor bases, namely 5-methylcytosine, N^6-methyladenine and N^4-methylcytosine are ubiquitous in prokaryotes [19-23]. Depending on the prokaryotic strain, one or two of these minor bases are nonrandomly distributed throughout the bacterial genome. Among the minor bases of DNA, N^4-methylcytosine has been predominantly found in thermophilic bacteria [23]. It has been proposed that N^4-methylcytosine replaces the thermally unstable 5-methylcytosine which would deaminate to thymine, giving rise to transition mutations. All minor bases arise postreplicatively by an enzymatic transfer of methyl groups from the cofactor *S*-adenosyl-L-methionine (AdoMet). The biological function of DNA methylation in prokaryotes has been partially elucidated. At least some of the methylated sites may be involved in directing mismatch repair systems to the correct strand of newly replicated DNA [24-27]. Other sites may belong to a host-specific restriction/modification system that obviously protects bacteria against viral infections by recognizing and destroying foreign DNA. Each restriction/modification system is composed of two functional elements: a) a restriction endonuclease that recognizes host-specific sites in the DNA and produces double-stranded cleavage unless these sites are fully methylated; and b) a modification enzyme (methylase) recognizing the same host-specific sites and protecting them against cleavage by methylation (reviewed in [20,28,29]). The prokaryotic restriction/modification systems are presently classified into the types I, II and III, depending on differences in enzyme structure and the underlying reaction mechanism (Table 1). The strain-specific recognition sites are either symmetric (type II), hyphenated and symmetric (type I) or totally asymmetric (type III). The type I and III enzymes are complex, multifunctional proteins which are able to both cleave and methylate DNA. They require ATP, AdoMet and magnesium for activity and cleave, possible randomly, at least 1,000 bp apart from the host-specific site (type I) or 25–27 bp to the 3′ end of the recognition site (type III). As one of the

Table 1. Characteristics of bacterial restriction and modification systems

	Type I	Type II	Type III
Restriction and modification activities	Single multifunctional enzyme	Two separate proteins; (i) *endo*nuclease, (ii) methylase	Single multifunctional enzyme
Protein structure	3 subunits	1 subunit	2 subunits
Genetic locus	hsdR, hsdM, hsdS	r and m	Res, mod
Sequence specificity	Hyphenated and symmetric	twofold symmetric	Asymmetric
Sites of cleavage	Possibly random, more than 1,000 bp from the recognition site	Inside the host-specific recognition site	25–27 bp 3′ to the host-specific recognition site
Requirements for methylation activity	AdoMet	AdoMet	AdoMet
Requirements for endonuclease activity	Mg^{2+}, ATP, AdoMet	Mg^{2+}	ATP

major differences between type I and type III enzymes, type I systems require AdoMet for restriction while type III systems do not [30]. Furthermore, in presence of ATP, unmodified DNA will be methylated and, in competition, restricted by type III enzymes. This is in sharp contrast to type I enzymes that either restrict or modify DNA, but never do both [31]. The type I restriction/modification systems are encoded by the host-specificity locus (*hsd*) that comprises three genes, *hsdR*, *hsdM*, and *hsdS*. The *hsdR* gene is responsible for restriction, *hsdM* for modification and *hsdS* for recognition of the host-specific site. As two promotors are present, translation of the transcripsts yields two proteins, the type I methylase (encoded by *hsdM* and *hsdS*) and the type I holoenzyme (encoded by *hsdR*, *hsdM*, and *hsdS*). Both the restriction enzyme and the methylase are able to methylate DNA, but only the holoenzyme is capable of restricting DNA. In contrast, type III systems are encoded by two genes, called *res* and *mod*. The *res* gene is responsible for restriction and the *mod* gene for both sequence recognition and modification. There are two promotors present, one in front of *mod* and the other preceding *res*, controlling the individual genes. The single *res* and *mod* polypeptides can be assembled either into the holoenzyme (endonuclease and methylase) composed of both subunits or into a methylase composed solely of the *mod* subunit. However, type III systems seem to be very rare — only three different type III systems have been described, EcoPI [32,33], EcoP15 [33,34], and HinfIII [35]. In contrast, type II systems are very common in bacteria. To date more than 400 have been identified and well characterized [36,37]. The type II enzymes appear to be structurally and functionally simpler than type I and type III enzymes. They require nothing but magnesium for activity, and cleave, in sharp contrast to the type I and type III enzymes, within the hostspecific recognition site or directly apart, leading to well-predictable cuts. Therefore, the type II restriction endonucleases have been proven as extremely useful tools in recombinant DNA technology. The related methylase is a separate protein that catalyzes transfer of a methyl group from AdoMet to either adenine or cytosine residues within the recognition sequence on each strand of the site.

As mentioned above, only a small amount of modified bases present in bacteria may belong to the restriction/modification system. In the well studied case of *E. coli* K-12, chromosomal DNA contains 0.5% N^6-methyladenine and 0.25% 5-methylcytosine [38], the reaction products of three different types of enzymes: a) an enzyme encoded by the *hsd* locus that methylates the second adenine in the AAC(NNNNNN)GTCG sequence, and belongs to the type I restriction/modification system [39]; b) an enzyme encoded by the *dam* gene, which methylates the adenine in GATC sequences [40]; and c) an enzyme encoded by the dcm gene, which methylates the internal cytosine in CCA/TGG sequences [41]. It is noteworthy that the adenines methylated by the *hsd*-coded enzyme account for less than 3% of the total level of N^6-methyladenines present in the DNA of *E. coli* K-12. Probably, only this minor fraction of N^6-methyladenines is involved in the restriction/modification system (type I) of this organism. However, the biological function of the major portion of the methylated bases in *E. coli* K-12 remains unknown. Some methylation sites may be involved in methyl-directed mismatch repair. For instance, the *E. coli* mismatch repair system is able to recognize noncomplementary base pairs in DNA and acts, apparently via localized excision and resynthesis, to replace mispaired bases [for review see [42]]. Probably the transient undermethylation of newly synthesized GATC sequences, localized in the region immediattely following the replication fork, directs the mismatch repair system to the newly synthesized strand, and thereby replication errors are removed [24,43,44]. Other methylation sites may be involved in recombination events [45,46], in protection of the DNA against nucleases [47,48], or in the regulation of gene expression [49].

Recently, two new restriction systems of *E. coli* K-12 have been described, *mcrA* and *mcrB* (for *m*odified *c*ytosine *r*estriction), which do not show related methylase activities, but attack DNA containing methylated cytosines in particular sequences [50].

3 DNA Methylation in Bacteriophages

Although viruses and bacteriophages are the simplest organisms and their genome contains a couple of genes only, several observations have been made that at least some viral genomes encode a virus-specific DNA methyltransferase. The function of this methyltransferase may be, at least in some cases, to protect viral DNA inside the infected host cell against restriction by host-encoded nucleases. In case of the T1 bacteriophages which infect *E. coli* strains, viral DNA is methylated at cytosine and adenine residues independently of the methylation capacity of the infected host cells [51]. For instance. the viral DNA is methylated at all GATC sites at adenine [52], even in *E. coli dam*-mutants that are incapable of methylating this sequence. This fact implies that the T1 phage encodes its own DNA methyltransferase, which has indeed been isolated recently [53]. Also other bacteriophages are capable of modifying DNA bases, sometimes in a particularly uncommon way. For example, the T-even phages incorporate 5-hydroxymethylcytosines into DNA, which are subsequently modified by glucosyl groups from UDP-glucose by phage-encoded enzymes [54]. The reason for this modification seems to be a protection against host-encoded restriction functions. Another unusual modification found in the *E. coli* bacteriophage Mu affects adenines in viral DNA which are converted to N^6-(1-acetamido)-adenines [55]. This modification is under the control of the virus-encoded *mom* gene. Also in this case, the modified DNA is resistant against a variety of host-controlled restriction nucleases [56].

4 DNA Methylation in Protista

In a variety of protista species, nuclear DNA contains either 5-methylcytosines or N^6-methyladenines as predominant minor bases [57,58]. On the other hand, two species of dinoflagellates have been described, belonging to the Zooxanthellaceae, which apparently lack 5-methylcytosines and N^6-methyladenines in their DNA, but contain considerably high amounts of the unusual base hydroxymethyluracil, arising by a specific modification of thymidine residues [58]. The biological role of this uncommon DNA modification is still obscure. Some Ciliata like *Paramecium* [59], *Oxytricha* [60], and *Tetrahymena* [61–64] obviously contain N^6-methyladenine as the only modified DNA base, located in AT sequences. Morphologically, most ciliate species contain two nuclei: a transciptionally silent diploid micronucleus and a transcriptionally active polyploid macronucleus. In *Paramecium aurelia*, both the macro- and the micronuclear DNA contains N^6-methyladenine (about 2,5 mol%), whereas the mitochondrial DNA from this organism is almost devoid of this modified base [59]. Also in other ciliata species N^6-methyladenine was found in the macro- and micronuclear DNA, ranging from 0.04 mol% to 2.5 mol% [61,65]. As no restriction/modification systems have been described in protista, the biological role of DNA methylation remains unknown.

5 DNA Methylation in Fungi

DNA methylation appears to be rare in the DNA of fungi. Antequera et al. [66] analyzed DNA isolated from mycelia of 20 species of fungi representing 15 taxonomic families. They found only 2 species which contained detectable cytosine methylation, as demonstrated by comparison of MspI and HpaII restriction patterns and nearest neighbour analysis. For instance, no 5-methylcytosine could be detected in mycelial DNA from *Aspergillus nidulans*, whereas *Sporotrichum dimorphosporum* and *Phycomyces blakesleeanus* had 0.2% and 0.5% of all cytosines methylated, respectively. The ascomycete *Aspergillus nidulans*, *Apodachlya* sp., and *Schizosaccharomyces pombe* virtually lack cytosine methylation [66,67]. Proffitt et al. [68] reported that less than 0.01% of cytosine residues were methylated in the DNA of yeast, *Saccharomyces cerevisae*. Bull & Wootton [69] reported that amplified DNA from *Neurospora crassa* mycelia was heavily methylated, whereas the nuclear DNA from the same organism had very low levels of DNA methylation. Jupe et al. [70] examined the fungus *Phymatotrichum omnivorum* and found in the DNA from the dormant sclerotia of this organism significantly more 5-methylcytosines than in the DNA from metabolically active mycelia of this fungus. However, nothing is yet known about the functional significance of DNA methylation in fungi.

6 DNA Methylation in Plants

In most lower plants, like green algae, DNA methylation levels are very low, except in *Chlorella* [71]. In contrast, DNA of higher plants is heavily methylated. In some species more than 30% of all cytosines are modified [72,73]. The majority of methylcytosines in plant DNA is found in the dinucleotide sequence CpG [74–76]. Additionally, a considerable amount of methylcytosines is found in the dinucleotide sequences CpA, CpT and CpC. Gruenbaum et al. [76] showed that methylation at these sites may be a consequence of methylating CpXpG trinucleotide sequences. Indeed, more than 80% of the trinucleotide sequences CpTpG and CpApG are methylated in wheat and tobacco, whereas non-symmetrical trinucleotides like CpApT are less than 4% methylated [76]. Despite these considerable differences to the situation in higher animals (see 7.2), not all methylatable sites are really methylated. Similar to animal DNA, approximately 60% to 80% of all CpGs are methylated in plant DNA. Some methylated regions may be critical for gene regulation (see 7.7), as, for example, in flax, where the least methylated regions in the transcriptional unit for ribosomal RNA correspond well with gene transcription [77]. Further evidence that DNA methylation may be involved in gene control in plants was provided by studies investigating the development of crown gall tumors in dicotyledonous plants, where parts of the Ti-plasmid, probably the disease-causing agent, were found unmethylated [78]. Whether this correlates with any changes in nuclear transcription patterns, remains to be examined. In order to assess the correlation between DNA methylation and gene expression in plants, Nick et al. [79] studied the DNA methylation pattern of the alcohol dehydrogenase gene from maize by genomic sequencing. They found that a 900 bp region 5′ to the transcriptional start site is not methylated in maize leaves, even though expression of this gene is totally repressed.

7 DNA Methylation in Animal Cells

A number of studies has shown that DNA methylation is species-specific. Although an adequate survey of the animal kingdom is still missing, the methylation pattern of DNA from various species can be catagorized into three different classes [80]: a) an arthropod pattern. b) an echinoderm pattern, and c) a vertebrate pattern. The basis for this classification is a difference in the overall level of DNA methylation, and the distribution of methylcytosines along the DNA [80]. Arthropod DNA almost completely lacks DNA methylation; vertebrate DNA has 1 to 6% of all cytosines methylated; and echinoderm DNA has intermediate levels of DNA methylation (0.2 to 1.5%). The arthropod genome, especially the insect genome, appears scarcely methylated [81–84], although notable exceptions have been reported [85]. In *Drosophila* some investigators even failed to detect any methylated bases [86]. The absence of DNA methylation in insect DNA is a particular problem for all the proposed functions of DNA methylation in higher animals, as insects are complex multicellular organisms that differentiate and develop in an orderly way with effective regulation of gene expression. Whether these organisms have developed mechanisms substituting the biological function of DNA methylation, or whether minute, yet undetectable, amounts of methylated bases are present in these insects that are sufficient to fulfil regulatory tasks, remains to be clarified.

7.1 The Methylation Reaction

There is no evidence that 5-methylcytosine can be incorporated into DNA via its triphosphate [87]. 5-methylcytosine in DNA arises by a postreplicative enzymatic reaction catalyzed by the enzyme DNA cytosine-5-methyltransferase (EC 2.1.1.37; DNA methyltransferase, DNA methylase). The methyl group donor is AdoMet. Most of the 5-methylcytosines in the DNA arise during the S phase of the cell cycle [88–90]. There is a lag of about one minute between synthesis of DNA and incorporation of methyl groups [91,92]. This is consistent with the finding that Okazaki fragments are ligated prior to methylation [93,94]. In addition to those methyl groups added in S phase, a significant portion of methyl groups is incorporated into the DNA several hours after DNA synthesis. This phenomenon is called delayed DNA methylation [95,96]. Presumably, methylation will be finished after completion of mitosis [97]. The majority of methyl groups, also those arising by delayed methylation, are incorporated into the DNA daughter strand, indicating that the methyl group transfers are predominantly maintenance methylation reactions preserving a pre-existing DNA methylation pattern in successive cell generations [98,99].

7.2 DNA Methylation Patterns

Methylated cytosines of vertebrate DNA are found in clusters localized in transcriptionally silent regions, heterochromatin, satellite DNA, and in repetitive sequences [100–103]. Unmethylated sequences of vertebrate DNA are also clustered and predominantly localized at the 5′ end of genes [10]. Echinoderm DNA contains large compartments of highly methylated DNA (m^+ compartment) and unmethylated DNA (m^- compartment). For instance, the m^+ compartment in sea urchins contains methylated sequences as long as 50 kbp that make up about 40% of the total genome. The rest of the genome belongs to the unmethylated compartment (m^-) and contains all actively transcribed regions. Interestingly, no redistribution of sequences between both compartments occurs during sea urchin development [104]. Similar m^+ and m^- compartments have been found in a variety of invertebrate species [80] and slime molds [105].

Several studies have concentrated on the sequence specificity of DNA methylation. In higher animals 5-methylcytosine is predominantly located in the dinucleotide sequence CpG [83,106–108], but is also detectable in small amounts in other cytosine-containing dinucleotides, i.e. CpC, CpT [109–110] and CpA [110]. In contrast to prokaryotes, both plant and animal DNA contains many potentially methylatable sites that remain unmodified, either because these sites are covered by proteins and are not available for DNA methyltransferases, or DNA methyltransferases possess a specificity towards the sequences surrounding the methylation sites. At least in *Neurospora crassa* evidence has been provided that the highly methylated zeta-eta chromosomal region contains a signal for DNA methylation [111]. However, although eukaryotic DNA methyltransferase appears unable to methylate each CpG sequence present in synthetic polymers [112,113], no additional sequence specificity has yet been identified in CpG flanking regions in vivo.

Repetitive sequences [101,103,114–116], satellite DNA [101,102,117–120] and heterochromatin [118,121] are the preferentially methylated regions of eukaryotic DNA, whereas euchromatin [118,121] and unique DNA sequences like coding regions [122] are less methylated. Tissue-specific and cell-specific differences in the extent of methylation of cytosines residues have been observed [122,123]. In humans, the most highly methylated DNAs originate from thymus and brain with 1 mole% 5-methylcytosine [122]. The two least methylated human DNAs are placenta and sperm DNA, which have 0.76 and 0.84 mole% 5-methylcytosine, respectively [122].

7.3 Distribution of CpG Sequences

The dinucleotide sequence CpG is the predominant methylation site in vertebrate DNA. On the other hand, CpG sequences are underrepresented, occurring at about one-fifth of the expected frequency [124–126]. The rarity of CpGs is probably caused by a failure of DNA repair mechanisms to recognize 5-methylcytosines deaminated to thymine [127,128]. The existence of such a mechanism was however demonstrated recently [129]. During evolution multiple CpG sites, formerly methylated, may have been converted to TpG and CpA by mutational events. This idea is stressed by the observation that concomitantly to the deficiency in CpG sites, genomes are enriched in TpG or CpA sequences [130]. Although 50 to 80% of all CpG dinucleotides in vertebrates are methylated, a certain part of all CpGs is organized in clusters and remains completely unmethylated [10]. In these unmethylated islands no CpG deficiency has been detected, probably due to absence of cytosine methylation. These clustered CpG dinucleotides located in G+C-rich regions have been designated HTF (Hpa II tiny fragments) islands, as HpaII restriction digests a part of vertebrate DNA to a heterogeneous fraction of short fragments [131]. HTF-islands are several hundred base pairs long and contain CpG at more than ten times the frequency of the total genomic DNA [132]. In spite of their common features concerning their undermethylation, different HTF islands do not contain related nucleotide sequences and are unique with respect to DNA reassociation [132]. The number of HTF islands may equate 30,000 resembling the anticipated number of genes (20,000–50,000). Indeed, evidence exists that HTF islands are gene-associated [133]. At least most housekeeping genes contain HTF islands, but some highly tissue-specific genes lack them. Most HTF-islands are located at the 5′ end of genes. They often include the first few exons as well as upstream sequences. Sequences flanking these islands are heavily methylated [10]. The biological function of HTF-islands remains unclear. There is evidence that transcription of genes with HTF islands is inhibited when the island is methylated. Many unexpected pseudogenes contain methylated islands. Furthermore, transfection experiments with an artificially methylated hamster adenosine phosphoribosyl transferase gene have demonstrated that methylation of the coding region of the gene does not influence transcription, whereas methylation of the 5′ region, which contains an HTF island, markedly reduces the level of gene expression [134]. Moreover, genes on the inactive X chromosome contain heavily methylated HTF-islands [135–137]. It is possible that methylation of HTF islands is a mechanism to inactivate no longer needed genes during evolution as Bird et al. [132] proposed.

7.4 Methylation of Organelle DNA

Methylation of mitochrondrial DNA has been found in a variety of vertebrate tissues and cell lines [138-139], but the levels are quite low, ranging between 0.2 and 1% of all cytosine residues. However, these findings have been criticized on the grounds of possible contamination of the mitochondrial preparations with nuclear DNA, since other investigators failed to detect 5-methylcytosine in mitochondrial material isolated from HeLa cells [140], in different organisms including yeast, *Neurospora*, calf and rat [141], and wheat [142]. In order to circumvent the problem of nuclear DNA contamination, Shmookler Reis & Goldstein [143] used mitochondrial-specific DNA probes to examine mitochondrial preparations from human diploid fibroblasts. They found minute amounts of methylcytosine in mitochondrial DNA. Pollack et al. [144] calculated that between 17 and 29 methylcytosines are localized throughout the total mitochondrial genome of mice. The biological function of mitochondrial DNA methylation is completely unknown. One hint may be presented by the observation that the level of mitochondrial DNA methylation decreases when cultured fibroblasts become senescent [143]. As substantial alterations of the mitochondrial genetic system is one of the first signs of senescent degeneration [145], it is possible that DNA methylation may be somehow involved (see 7.14).

Chloroplast DNA of higher plants seems devoid of methylcytosines [62,146,147]. In contrast, in the chloroplast DNA of female gametes and zygotes of the green algae *Chlamydomonas*, 5-methylcytosine is easily detectable, whereas male gametes and vegetative life forms of this organism apparently lack DNA methylation [148].

7.5 DNA Methylating Enzymes

DNA methyltransferases from bacterial cells have been purified and extensively characterized in numerous systems (see Chapt. 2). Due to low abundance of this enzyme in animal cells, the purification of DNA methyltransferases from mammalian cells has been difficult. The first reports dealing with the characterization of a mammalian DNA methyltransferase originate from 1967 and 1968 [149-151]. Subsequently, DNA methyltransferases have been purified from a number of different sources. The assay system for the determination of DNA methyltransferase activity is based on the incorporation of [^{3}H] methyl groups from labeled AdoMet into the DNA substrate. The DNA is then isolated for determination of radioactivity by proteinase and RNase treatments in order to remove methylated proteins and RNA. The methylated product of the enzymatic reaction was found to be exclusively 5-methylcytosine. The methyl-accepting ability of a particular substrate depends on its G+C content. Thus, substrates like poly (dG—dC). poly(dG—dC) or the 72% G+C-rich *Micrococcus luteus* DNA are good methyl acceptors. Surprisingly, however, the synthetic polymer poly(dI—dC)-poly(dI—dC) shows the highest methyl-accepting capacity among all substrates tested [152,153]. The reason for this is unknown, but it is assumed that the structurally less tight DNA polymer poly(dI—dC) · poly(dI—dC) is more easily accessable for the DNA methyltransferase enzyme. Mammalian DNA methyltransferases are able to methylate the DNA from the same cells from which the enzyme has been isolated [154-158]. This finding may be explained by the fact that some methylatable sites are covered within the cell by chromosomal proteins, which prevent those sites from becoming methylated in vivo.

The binding of DNA methyltransferase to the DNA substrate is probably random and independent of sequences recognized by the enzyme as methylation sites [159–161]. The formation of enzyme DNA-complexes was shown to be independent of the presence of the cofactor AdoMet [162]. The DNA methyltransferase, once bound to DNA, accomplishes as series of methyl group transfers without being detached from the DNA polymer [157,158,162–164]. This mode of action can be described as a "walking mechanism" [163]. DNA methyltransferase binds preferentially to single-stranded DNA regions [159,165]. It has been proposed that binding of DNA methyltransferase to native DNA may involve "breathing" of the DNA allowing the formation of a tight enzyme-DNA complex [159,166].

Two types of enzymatic activity have been postulated for DNA methyltransferases: a maintenance type and a de novo type of activity [4,5]. The maintenance activity is responsible for the semiconservative copying of the DNA methylation pattern and the de novo activity allows the alteration of a preexisting methylation pattern by transferring methyl groups to previously unmethylated sites. It was soon recognized that the mammalian DNA methylating enzymes prefer substrates containing hemimethylated sites over unmethylated DNAs. These substrates are thought to mimic the situation at the replication fork, where the DNA polymerase has left a hemimethylated site after incorporation of a dCMP residue in the sequence CpG. Hemimethylated substrates were prepared from ethionine-treated [167] or methionine-starved cells [166], and from cells treated with the methylation inhibitor 5-azacytidine [158,168]. When cells are incubated for one cell cycle in presence of this agent, the cytosine analogue 5-azacytidine is incorporated into the DNA and prevents methylation of the daughter strand. This inhibition of maintenance DNA methyltransferase is strong [158,168,169] and some authors think that a covalent binding is involved [170]. The in vivo prepared hemimethylated 5-azacytosine-containing DNA is a good substrate for DNA methyltransferase in vitro. Other hemimethylated DNA substrates were prepared by nick translation of DNA in presence of 5-methyl-dCTP [160,165,171] or by primer extension of single-stranded phage DNA in presence of 5-methyl dCTP [172–174]. With any of these substrates a strong preference of DNA methyltransferase for the hemimethylated DNA has been observed. Depending on the particular DNA substrates used, the rate of methylation of hemimethylated ones is about 5 to 100-fold higher than the rate of methylation of de novo substrates. Therefore, it can be postulated that the mammalian enzymes are primarily maintenance DNA methyltransferases. However, they possess a considerable and significant de novo DNA methyltransferase activity. Due to the relative stability of the DNA methylation patterns in vivo [175], it must be postulated that this de novo activity is largely suppressed within the cell and is allowed to occur only in certain situations of development [176]. How such a mechanism, allowing only maintenance methylation, might work, is completely unknown at present.

The sequence specificity of DNA methyltransferase in vitro reflects the distribution of methyl groups along the DNA in vivo. Sequence specificity analysis has been carried out either by nearest neighbor analysis [155,172,177,178], or by separation of dinucleotides [179] and by restriction enzyme analysis [165,171,179,180]. The predominant methylation site is the CpG dinucleotide. However, methyl groups are also incorporated by the enzyme into the dinucleotide sequences CpC [165,177,178], CpA [177,179] and CpT [177,179]. Occurrence of methylated cytosines outside of the CpG sequence has been also found in vivo (see 7.2).

In addition to this surprisingly limited sequence specificity, DNA methyltransferase seems to respond to secondary structures in DNA. As already mentioned, the DNA polymer poly(dI—dC) · poly(dI—dC) that is structurally different from poly(dG—dC) · poly(dG—dC), is a preferred methyl-accepting substrate by DNA methylase. Covalently closed circular DNA in the left-handed Z-form has been shown to be resistant to methylation by mouse DNA methyltransferase [181], whereas linear Z DNA is a substrate for DNA methyltransferase from human placenta [158]. Significantly different results were obtained with linear in comparison to circular Z DNA substrates using bacterial DNA methyltransferases [182]. The sequence specificity was shown to be dependent on the supercoiling of the substrate DNA as was demonstrated by a differential distribution of incorporated methyl groups in a plasmid genome [181]. Furthermore, the enzyme selectively methylates DNA sequences containing mispairs [183,184]. The enzyme seems to be highly sensitive to the recognition of any single-stranded DNA region. At present, it is not completely clear, whether every CpG dinucleotide, the predominant methylation sequence, can become methylated in vitro. Complete methylation has been obtained with a rat DNA methyltransferase [179], however not with mouse [165,180] or with human enzymes [171]. It was shown that the primary DNA sequence is important for the selection of those CpG dinucleotides which are methylated by DNA methyltransferase and those which are not. The presence of at least two CpG dinucleotides and a spacing of 13 to 17 nucleotides between them seems to be required for maximal methylation of the 3′ CpG site, whereas the 5′ CpG sequence itself is not methylated [112,113]. Additional investigations of this kind are certainly desirable in order to gain more insight into the DNA sequence dependent specificity of DNA methyltransferase.

The specificity of DNA methyltransferase may be modulated by the presence of DNA binding proteins in vivo. It was shown that DNA methyltransferase activity is diminished in the presence of chromatin proteins, mainly histones [185]. Recently we have identified two different non-histone protein components which are able to increase the activity and change the sequence specificity of DNA methyltransferase on its substrate DNA (Tomassetti et al., in preparation). It is conceivable that the DNA methylation pattern is established in the cell in concert with numerous other DNA-protein interactions, some of which may inhibit while others may stimulate the methylation of a particular sequence.

The polypeptide composition of mammalian DNA methyltransferase has been elucidated. Molecular weight determinations using sedimentation analysis or gel filtration indicated a molecular weight of this enzyme above 100,000 [155,157,158,162,174,177]. The enzyme seems to be present in a monomeric form; only one report postulates that the human DNA methyltransferase is a dimer [186]. Homogeneous preparations of mouse DNA methyltransferase have been obtained, which indicate that the native DNA methyltransferase is a polypeptide of 190 kDa [160,165,178,187]. This polypeptide appears to be extremely susceptible to proteolytic degradation and this degradation results in polypeptides of molecular masses 150–180 kDa, and 100–110 kDa [178,188]. The situation appears to be complicated in mouse Friend erythroleukemia cells, since the appearance of some of these polypeptides of lower molecular weight was shown to be dependent on the proliferative state of the cells [187]. Preparations of homogeneous DNA methyltransferase were obtained by conventional chromatographic procedures [164,178,187] or by immunoaffinity chromatography using monoclonal antibodies directed against human DNA methyltransferase [160]. These monoclonal antibodies were shown to cross-react with rodent DNA methyltransferase [160,165], which indicates a certain evolutionary conversion of DNA methyltransferase structure. Anti-DNA methyltransferase antibodies were used to immunoprecipitate DNA methyltransferase polypeptides of 190 kDa also from human cells [188,189] indicating a similar size of mouse and human DNA methyltransferase. DNA methyltransferase was also purified from human placenta by immunoaffinity chromatography [160]. Polypeptides of 158, 150, 110 and 50–68 kDa were obtained and it was speculated that these polypeptides of lower molecular weight were a result of partical proteolytic cleavage. The most abundant DNA methyltransferase polypeptide in human placenta was that of 110 kDa. A polypeptide of similar size was purified by other methods from terminal human placenta [174]. It remains to be seem whether the smaller size of DNA methyltransferase obtained from human placenta is an artifact of preparation due to the high endogeneous proteolytic activity in this tissue or whether it represents a real in vivo situation with different DNA methyltransferases present in this extraembryonic tissue. In this connection it should be mentioned that DNA from extraembryonic tissues contains much less 5mC than DNA from other somatic tissues [190–193]. A posttranslational modification of DNA methyltransferase may be of importance for regulation activity of the enzyme. The DNA methyltransferase polypeptide is phosphorylated in vivo (our unpublished data) and is a preferred substrate for protein kinase C in vitro [194]. However, this phosphorylation seems to have only small effects on the enzyme activity. All DNA methyltransferases purified from mammalian cells prefer hemimethylated over unmethylated DNAs and this also applies to those enzymes purified to apparent homogeneity. This is consistent with the idea that one enzyme molecule carries out both types of these activities as it has been previously postulated [158,162].

DNA methyltransferase has been commonly isolated from nuclei, from which in most cases more DNA methyltransferase activity was recovered than from cytoplasmic fractions. Thus, the enzyme's localization was inferred to be predominantly nuclear. This was confirmed by immunocytochemistry [189]. The DNA methyltransferase antigen shows a granular staining pattern with the nuclear regions clearly dominating. Most of the enzyme is readily extractable in presence of buffers containing 0.3 M salt. The bulk of DNA methyltransferase seems to be bound to linker DNA in condensed regions of chromatin [195]. A considerable amount of DNA methyltransferase is localized in a cellular compartment, which is defined as the nuclear matrix represents those nuclear structures which remain insoluble in 2 M NaCl and are resistant to DNase I digestion. The presence of DNA methyltransferase in the nuclear matrix was demonstrated by biochemical [196] and by immunocytochemical [189,197] techniques. The relative amount of DNA methyltransferase within these structures is difficult to quantify, but may not exceed 20% of total DNA methyltransferase activity. However, since the nuclear matrix is thought to be the site of the DNA replication process, the presence of DNA methyltransferase may indicate its functional role in the maintenance of the DNA methylation pattern. Accordingly,

DNA methyltransferase activity has been detected in complexes of replication dependent enzymes, the so-called replicase complexes [198,199].

It has long been recognized that proliferating tissues and cells are enriched in DNA methyltransferase activity as compared to non-proliferating ones. Hence, DNA methyltransferase has been preferentially isolated from proliferating cells or tissues. The number of extractable DNA methyltransferase molecules was estimated to be 20,000 to 40,000 in proliferating human or mouse cells [189]. The presence of the enzyme clearly depends on the proliferative state of the cells. In stationary phase cells, the amount of DNA methyltransferase is drastically reduced [188]. When non-proliferating cells are subjected to a mitogenic stimulus, the DNA methyltransferase levels increase significantly. This was demonstrated in hepatectomized rat livers [200,201], where the activity increases 2- to 3-fold after hepatectomy. Similarly, the DNA methyltransferase levels increase at least 20-fold from near zero levels after stimulation of peripheral lymphocytes with lectins [189,201,202]. A cell-cycle-dependent regulation of DNA methyltransferase activity is certainly important for the stability of DNA methylation patterns in absence of DNA synthesis. The low levels of DNA methyltransferase found in non-proliferating cells may be involved in DNA repair processes. Methylation of DNA repair patches was shown to be more effective in proliferating than in non-proliferating cells [203]. This may suggest that the "replicative" DNA methyltransferase is also involved in methylation of repaired regions, although it cannot be excluded at present that different DNA methyltransferases participate in these two cellular functions.

The frog iridovirus 3 contains the most highly methylated DNA of all known animal viruses with more than 20% of all cytosine residues methylated by a virus-encoded DNA methyltransferase [204]. In order to allow transcription of the highly methylated viral genome by host RNA polymerases, the viral genome encodes for a *trans*-activating factor which modifies RNA polymerase II of the host at an early stage of transcription [205]. However, the reason for the high methylation of the frog virus genome is obscure.

In plants, the preferred methylated DNA sequence is CpXpG (see 6). The DNA methyltransferase responsible for the methylation of these sequences appears to be significantly different from the mammalian enzymes. The enzyme isolated from germinating wheat embryos has a molecular weight of 50,000 and transfers methyl groups to both unmethylated and hemimethylated substrates to about the same extent [206]. The preferred substrate is endogenous, double-stranded DNA. The wheat enzyme resembles the DNA methyltransferase from *Chlamydomonas reinhardii* [207], suggesting that plant cells possess their own type of DNA methyltransferase.

7.6 Formation of Chromatin Structures

It is widely accepted that there is a relation between chromatin structure and DNA methylation. DNA in transcriptionally active regions of chromatin is hypomethylated compared to DNA in inactive regions. In fact, only one-third of all CpG dinucleotides located in active regions of chick erythrocyte DNA are methylated compared to two-thirds in the bulk DNA [208]. When actively transcribed DNA from the same source, selected by hybridisation to polyadenylated messenger RNA, was analyzed, this complementary DNA had only 25% of its CpG residues methylated as compared to 65% for the total DNA [208]. These data were obtained by nearest neighbor analysis and Davie and Saunders [209] were able to confirm them by investigating mouse L929 cells and calf thymus DNA by HPLC analysis. Furthermore, by use of methyl cytosine-specific antibodies Ball et al. [210] confirmed that transcriptionally active chromatin regions, enriched in high-mobility group proteins (HMG), are more than 2-fold undermethylated in comparison to bulk DNA, and, on the other hand, more than 80% of all methylcytosines were detected in transcriptionally inactive regions. Considering the non-random distribution of 5-methylcytosine between nucleosome core DNA and linker DNA [100], methylation of eukaryotic DNA may be important for the condensation of the DNA into chromatin and for the positioning of nucleosomes and the coiling of the nucleosome chain to the final chromatin structure. Thus, the high degree of methylation of distinct parts of the genome may contribute to the packaging of genes into highly condensed chromatin which renders them inaccessible to activation.

With regard to the interdependence of a specific DNA methylation pattern and particular chromatin configurations, the key question is, whether the chromatin structure is established according to a preexisting DNA methylation pattern or whether the methylation pattern results from DNA binding proteins excluding the access of the DNA methylating enzyme to the possible methylation sites. The latter is certainly the case in vitro, since chromatin proteins were shown to inhibit DNA methyltransferase. The methyl-accepting ability of isolated nuclei is increased after trypsinization or salt extraction of chromosomal proteins [185]

The effect of DNA methylation on chromatin structure has been studied by DNA-mediated gene transfer experiments. Transfected genes are frequently found in an active chromatin configuration and are expressed, since this is often a consequence of the selection process. A correlation between undermethylation of DNA, presence of DNase I hypersensitivity and gene expression was observed [reviewed in [211]]. The real molecular basis explaining the nature of these nuclease hypersensitive sites is unknown but it is likely that an alteration of the composition of DNA binding proteins in these regions is involved. In order to test whether DNA methylation is causatively involved in gene repression, completely methylated gene constructs were inserted into mouse L-cells [212]. Methylation prevented the formation of nuclease sensitive chromatin structures, whereas the corresponding unmethylated construct was present in an active chromatin conformation. Transcription of the herpes simplex virus thymidine kinase gene can be blocked by DNA methylation. However, this is a late and indirect effect, which requires the formation of chromatin [213]. In vitro reconstituted chromatin on methylated and mock-methylated DNA was microinjected into recipient cells. The chromatin was found to be inactive only after prior methylation of all HpaII sites. In contrast the mock-methylated construct was fully active. The inactivation was reversible after treatment of cells with 5-azacytidine [213]. The molecular difference of chromatin on methylated and unmethylated DNA regions remains unclear. A DNA binding protein being a possible candidate for methylation-mediated formation of chromatin structures has been isolated from human placenta, which binds preferrentially to 5-methylcytosine-rich DNA [214]. This protein was shown not only to be methylation-specific but also sequence-specific [215]. On the other hand, the transcription factor Sp1, the binding sequence of which contains a CpG-dinucleotide and is present at the 5′-end of several genes, does not discriminate between methylated and unmethylated sequences [216].

Methylation of cytosines may not only have a direct effect on the formation of chromatin structures. Moreover, this effect may be mediated through DNA structural alterations. DNA methylation favours the formation of left-handed *Z* DNA [217–219]. *Z* DNA formation in vivo may be recognized by specific DNA binding proteins with possible consequences for regulation of gene expression. However, it remains to be elucidated whether methylation of alternating CG stretches brings about such effects in vivo. DNA methylation may induce other structural alterations of the polynucleotide chain, e.g. DNA curvature [220].

7.7 DNA Methylation and Gene Expression

DNA methylation has been implicated in the regulation of gene expression in eukaryotic cells [4,5,221]. Supporting experimental evidence has been provided by three different outcomes: a) a positive correlation between undermethylation and expression of eukaryotic and viral genes in vivo; b) the induction of gene expression by DNA demethylating agents; c) the suppression of microinjected genes methylated in vitro prior to transfection.

An inverse correlation between the level of DNA methylation and transcriptional activity has been described for a variety of tissue-specific eukaryotic genes, including the albumin gene [222–225], the α-fetoprotein gene [225–227], the prolactin and growth hormone gene [228], the metallothioneine-I gene [229], the gamma-interferon gene [230], the Thy-1 gene [231], the zona pellucida ZP-3 gene [232], the chicken cardiac myosine light chain gene [233], the rat kappa-casein gene [234], the human thyroglobulin gene [235], and the chicken vitellogenin gene [236,237]. For housekeeping genes, e.g., the adenine phosphoribosyltransferase gene and the dihydrofolate reductase gene [238], as well as the phosphoenolpyruvate carboxykinase gene [239], an inverse correlation between DNA methylation and gene expression was found. However, for a number of other genes no obvious correlation between DNA methylation and gene expression has been observed in vivo. For example, during development of chicken or *Xenopus*, the hormone inducible vitellogenin genes [240,241] or the δ-crystallin gene [242] became expressed despite being highly methylated. Other examples are the α_2(I) collagen gene [243] or X chromosome-linked genes in extraembryonic tissues [244], where hypomethylation of DNA was not correlated with gene expression. It is important to emphasize that all such experiments were carried out by use of restriction enzymes to determine the methylation status of the respective genes. As the available enzymes recognize only 10–15% of the total CpGs present in DNA and do not distinguish between hemi- and fully methylated sites, the significance of these results must be carefully considered. Possibly all those experiments, which showed no correlation between expression and methylation, have failed to detect the functionally relevant sites. To prevent the disadvantage of restriction enzymes, a new method, genomic sequencing, has been developed, which allows the direct assessment of single methylation sites by discriminating the plus and minus strand [245–247]. Using this method, it could be demonstrated that fully demethylated DNA strands do not appear to be an essential prerequisite for activation of the vitellogenin gene [247].

These authors provided evidence that strand-specific demethylation of particular sequençes at the promotor region are also rendering the gene transcriptionally active in the hemimethylated state.

One of the best investigated systems are hemoglobin genes. Hemoglobin genes of mammals and birds consist of two clusters of genes, namely the α gene cluster and the β gene cluster [248]. The genes within these two clusters probably evolved by duplication of an ancestral globin gene and diverged considerably during evolution. The individual genes of globin clusters become consecutively expressed during embryonic and fetal development (5′ to 3′). It is well established that tissue-specific methylation patterns exist for the globin genes of mammals and birds, and an inverse correlation between DNA methylation of particular HpaII/MspI sites at both 5′ and 3′ sides of globin genes and gene expression was described [249–254]. Furthermore, Mavilio et al. [255], who investigated the selective expression of embryonic, fetal and adult β-globin genes in human yolk sac, fetal liver and adult bone marrow, found that a strong correlation between DNA hypomethylation in the flanking sequences of globin genes and their expression exists during all phases of ontogenic development. They concluded that DNA methylation may represent a key mechanism for the developmental regulation of globin gene switching, from embryonic to fetal, and from fetal to adult. These observations were extended by Busslinger et al. [256], who demonstrated that in vitro methylation of cytosine residues near the 5′ end of the human gamma globin gene prevented expression of this gene in mouse L-cells, whereas methylation of the coding region had no effect on gene expression. In a first approximation, an upstream region between —760 and +100 was found to be important for methylation-associated gene suppression. A recent work by Murray and Grosfeld [257] defined the boundaries of this region in more details. They demonstrated that methylation downstream of +92 does not affect transcription, while methylation downstream of —210 abolishes transcription completely. Furthermore, Murray & Grosveld reported that removal of methylated residues either 5′ or 3′ to the cap site did not alleviate suppression, whereas removal of both 5′ and 3′ methylated residues did partially restore globin gene expression [257]. Further evidence that DNA methylation may be involved in globin gene expression was published by Ley et al. [258], who treated the mouse erythroleukemia hybrid cell line M11-X with 5-azacytidine. The M11-X line contains most of the human chromosome 11, including all human β-globin genes, but does not express any embryonic or fetal human hemoglobin. Treatment of the M11-X cells with 5-azacytidine resulted in the selective expression of the human gamma globin gene. However, restriction analysis demonstrated that the 24 h exposure of M11-X cells to 5-azacytidine caused hypomethylation of both fetal and embryonic globin genes [258]. Until now the reason why embryonic genes remained suppressed is unknown.

For immunoglobin genes it has been demonstrated that demethylation accompanies class switching of heavy chains [259], somatic rearrangement of light chain genes [260], and expression of both successively rearranged heavy and light chain genes [261].

In case of the albumin gene, Ott et al. [222] identified a critical MspI site for albumin gene expression 80 base pairs downstream from the 5′ end of the first exon, whose undermethylation appeared to be necessary but not sufficient for stable albumin gene expression in rat hepatoma cells. In somatic rat hepatoma — mouse L cell/

fibroblast hybrids, not expressing the albumin gene, the MspI site of the albumin gene remained unmethylated. This implies that other factors may be responsible for the efficient suppression of the albumin gene in this system [262)]. These results are in disagreement with data from Orlofsky and Chasin [263)], who reported that non-producer variants deriving from a high producer Fu5-5 hepatoma cell line, which do not show any albumin expression, underwent de novo methylation of an 8-kilobase region, with boundaries in the 5′ flanking and in the middle of the gene. Six independent hybrid clones, obtained by crossing a high producer line with a null variant, showed extinction of albumin production and concomitant hypermethylation of the 8-kilobase domain. Interestingly these hybrids retained the capacity for de novo methylation of these specific sites [263)].

In contrast, a number of other genes have been analyzed where no obvious correlation between DNA methylation and gene expression has been found. For example, the DNA around the start site of transcription of the α_2(I) collagen gene in chicken is undermethylated, and the central and 3′ regions are methylated, irrespective of transcription [243, 264)]. Similar results were obtained for the type II and type I collagen genes in differentiated and dedifferentiated chondrocytes [265)], and for two rat insulin genes and the rat α-fetoprotein gene [266, 267)]. Also Nelson et al. [268)] reported that lipopolysaccharide-induced transcription of the kappa immunoglobin gene occurs on both allels independent of the methylation status. An imperfect correlation between gene expression and undermethylation was also described by Hjelle et al. [269)] for the human growth hormone and chorionic somatomammotropin genes in expressing and non-expressing tissues.

A particularly conclusive evidence that DNA methylation at specific sites is causally related to gene inactivation came from microinjection experiments using either *Xenopus* oocytes or mammalian cells as recipients for unmethylated or in vitro methylated cloned DNA of viral or eukaryotic origin [270–272)]. These experiments demonstrated that a variety of genes can be efficiently inactivated by methylation of specific sequences, mainly at the 5′ end of the transferred genes. For in vitro methylation either prokaryotic HpaII DNA methyltransferase [270–272)] or eukaryotic DNA methyltransferase [179)] was used. During the last decade several procedures have become available that allow the introduction of foreign DNA into eukaryotic cells. One possibility of inducing a cell to pick up added DNA is to adsorb this DNA onto calcium phosphate particles [273)]. The incorporated DNA will integrate at random sites into the recipient cell's chromosoms. As the transfection efficiency of this procedure is very low, cotransfection of two genes, a) the gene of choice and b) a gene expressing a selectable marker, is usually performed in order to identify recombinant cells. Cotransfection of tk^- cells can be performed with the thymidine kinase gene (tk gene) and recombinant cells can be easily detected under selective growth conditions in HAT medium (medium containing hypoxanthine, aminopterine, and thymidine). Instead of calcium phosphate precipitation, DNA can be injected directly into recipient cells by microinjection. The DNA used in these transfection experiments has usually been propagated in bacterial hosts and therefore lacks its normal methylation pattern. When such unmethylated DNA is introduced into vertebrates cells, it remains largely unmethylated for many generations due to an effective suppression of de novo methylation in most eukaryotic cells and tissues [175, 274, 275)]. The situation is quite different when in vitro methylated DNA is transfected.

In this case the fate of methylcytosines depends on the integration site into the recipient genome. Wigler at al. [175] transfected mouse cells lacking a functional thymidine kinase (tk) gene with either methylated or unmethylated herpes simplex virus or chicken tk genes. They found that the unmethylated tk gene sequences revert tk^- cells more efficiently than the methylated ones. Many other similar studies have been performed introducing a variety of vertebrate and viral genes, including the hamster adenine phosphoribosyltransferase gene [272], herpes simples virus thymidine kinase gene [274, 276], late genes from SV 40 [277], adenovirus 2 genes [270, 278], mouse mammary tumor virus genes [279], and Moloney murine leukemia virus genes [280, 281]. In all these experiments transforming efficiency and gene expression of unmethylated DNA was much higher than that of methylated DNA. In contrast, two examples have been reported where methylation of all CpGs failed to have biological significance regarding gene expression: a) SV40 DNA was methylated in vitro at all CpG sites by rat liver methyltransferase and microinjected into cells. It was still possible to transcribe the genes efficiently from the early region [282]; b) ribosomal DNA of *Xenopus* was found fully methylated, but nevertheless actively transcribed when injected into *Xenopus* oocytes [283]. These observations suggest that expression of certain genes may be insensitive to inhibition by DNA methylation.

As gene expression is regularly altered during development, several studies have been initiated to measure developmental variations of DNA methylation patterns of particular genes [239, 242, 253, 284–287]. Young & Tilghman [288] demonstrated that a genome-wide loss of 5-methylcytosine occurs in induced, silent, and constitutively expressed genes when mouse F9 teratocarcimoma cells are induced to differentiate. Grainger et al. [242] investigated methylation of the chicken lens δ-crystallin II gene during early embryogenesis and found that some hypomethylation events concurred with the activation of the gene, but it was not clear whether this hypomethylation precedes gene activation. Benvenisty et al. [239] analyzed the extent of methylation of 18 HpaII and HhaI sites in the phosphoenolpyruvate carboxykinase (PEPCK) gene in various fetal and adult tissues. This analysis revealed that distinct methylation patterns exist which correspond to the transcriptional activity of this gene.

Most studies analyzing DNA methylation pattern of viral DNA have also revealed a good correlation between undermethylation and gene expression. DNA isolated from intact eukaryotic virus particles (adenovirus and baculovirus) does not contain methylcytosines [289]. In contrast, when viral DNA is integrated into the host genome, it becomes methylated. Especially non-transcribed regions coding for virion proteins are heavily methylated [290–292]. Later in infection, when viral DNA or adjacent cellular sequences are encapsidated in virions, they lose their methylcytosines completely [293]. Kruzcek and Doerfler [294] have precisely measured the methylation state of HpaII/MspI and HhaI sites in three adenovirus 12-transformed hamster cell lines. In all cases where early genes were transcribed into mRNA, CpG sites located at the promotor-leader sequence and at the 5′ end of the gene were undermethylated in comparison to unexpressed early genes. Sites within the coding region of early genes may or may not be methylated. Non-expressed early genes are usually, but not always, methylated, but all late genes are entirely methylated. In some adeno virus 12-transformed rat brain cells late genes are expressed and, in such cases, undermethylated [292]. Accordingly, Guntaka et al. [295] provided evidence that methylation of the enhancer region in the domain between nucleotides − 208 and −119 of the U3 region in the long terminal repeats of avian sarcoma virus suppresses viral gene transcription. Also for the chicken retroviral loci ev-1 and ev-3 it was shown that a good correlation exists between undermethylation and gene transcription [296]. Moreover, many other viruses have been described which demonstrate a negative correlation between viral gene expression and DNA methylation, including avian sarcoma virus [297], mammary tumor viruses [298], herpes simplex virus [299], herpes saimiri virus [290], mouse endogeneous type C virus [300], Epstein-Barr virus [301], Marek's disease virus [302], and AKR murine leukemia virus [303]. Thus, high methylation levels of viral DNA may be present in latently infected cells, whereas methylation of the virus DNA sequences is not detectable in productively infected cells. However, two notable exceptions are the murine Moloney leukemia virus in teratocarcinoma cells [304] and the herpes simplex virus in the central nervous system of latently infected mice [305]. Here, the viral genes are not expressed although they are undermethylated.

DNA demethylating agents like 5-azacytidine can induce expression of a large variety of genes, including for example the metallothionein I gene [306], the herpes simplex thymidine kinase gene [307], the fetal hemoglobin genes [308–310], the prolactin gene [311], and the glucocorticoid receptor gene [312]. Genes can also be activated by 5-azacytidine when they are located on the inactive X chromosome [313–315]. Although it is well established that 5-azacytidine causes a tremendous undermethylation of DNA in vivo and in vitro, it is important to prove whether this genomewide undermethylation is somehow related to gene expression. Evidence for this fact derives from the observation that 5-azacytidine-related changes in gene expression are heritable for many generations even in absence of this drug [311,313,316,317]. Furthermore, 5-azacytidine-related gene expression is accompanied by hypomethylation of particular sites within the expressed genes [296,306,307,310]. Since gene activation frequencies under 5-azacytidine treatment are several orders of magnitude higher than mutation frequencies [229,314,317,318], epigenetic mechanisms like DNA methylation may be causally involved.

Rösl et al. [319] reported that the transcription of human papilloma virus (HPV) type 18 is selectively suppressed in non-tumorigenic HeLa × fibroblast or HeLa × keratinocyte cell hybrids by 5-azacytidine. In tumorigenic hybrids and in the parental HeLa cells viral gene expression is not suppressed by 5-azacytidine. Down regulation of HPV 18 mRNA in hybrids correlates directly with cessation of cellular growth and can be abolished by cycloheximide. Rösl et al. postulated an intracellularly acting control mechanism, directed against papillomavirus transcription, which is induced by 5-azacytidine and seems to depend on the presence of specific chromosomes in non-tumorigenic cells [319].

Although the evidence strongly suggests that DNA methylation has a role in gene regulation, it is not yet known whether DNA methylation is involved in the primary events abolishing gene expression, or whether DNA methylation is a secondary event that occurs on already inactivated genes, perhaps in order to solidify the state of transcriptional silence.

7.8 X-Chromosome Inactivation

In female mammalian cells, one of the two X chromosomes is transcriptionally silent, thus allowing a gene dosage compensation of X-linked genes in male and female cells. This X chromosome inactivation, which takes place in early embryonic development, is, according to present thinking, a three step event: 1) one of the two X chromosomes is selected for inactivation and the inactivation process is initiated; 2) the inactivation spreads over the whole chromosome and 3) a stable and heritable repression of the genes on this chromosome is achieved [320]. A later reactivation occurs in vivo only in germ line cells prior to meiosis [321]. The inactivation of one X chromosome does not occur simultaneously and in the same way in all cells during development. The event occurs first in extraembryonic cells, where the X chromosome of paternal origin is preferentially inactivated in rodents [322, 323], whereas either paternal or maternal X chromosomes are selected at random for inactivation in humans [324]. Later, in the embryo proper of all eutherian mammals, X chromosomes of both paternal and maternal origin are selected again randomly for inactivation [325]. Once inactivated, the silent state of this chromosome is inherited throughout the daughter cell generations and the genes on that chromosomes remain silent in most tissues. Considerable species differences exist, inasmuch as in marsupials, similar to the situation in extraembryonic tissues of rodents, the inactivated X chromosome is exclusively of paternal origin [326].

In recent years, a large body of evidence has accumulated showing that enzymatic DNA methylation is involved in X chromosome inactivation in eutherian mammals. Most results point to the possibility that DNA methylation is involved in the above mentioned point 3, the maintenance of X chromosome inactivation. This maintenance is guaranteed by the clonal stability of the DNA methylation pattern [175, 274, 275]. The evidence for an involvement of DNA methylation in X chromosome inactivation can be summarized as follows: 1) Gene transfer experiments have shown that hypoxanthine phosphoribosyl transferase (HPRT) sequences derived from an inactive X chromosome are relatively inefficient in transferring the $HPRT^+$ phenotype to $HPRT^-$ recipient cells, whilst the DNA from an active X chromosome is efficient in DNA-mediated gene transfer [327]. This means that the DNA from the inactive X chromosome must be modified. 2) Treatment of cells with 5-azacytidine, a strong inhibitor of DNA methyltransferase, leads to reactivation of genes on the inactive X chromosome. HPRT sequences from 5-azacytidine-reactivated cells function normally in DNA-mediated gene transfer [313, 328]. 3) Differential methylation patterns were found in particular regions of genes on active and inactive X chromosome, although no global difference of DNA methylation seems to exist between an active and an inactive X chromosome, as probed by using random X chromosome derived sequences [329, 330]. The differentially methylated regions represent promotor and enhancer sequences. They were found to be consistantly hypermethylated on inactive X chromosomes. This correlation was established for the following genes: glucose-6-phosphate dehydrogenase [331]; hypoxanthine phosphoribosyl transferase of humans [136, 137] and mouse [332], phosphoglycerate kinase [333, 334] and genes of unknown functions located 3′ to the glucose-6 phosphate dehydrogenase locus [135, 335]. It is important to note that this differential degree of methylation was found only in CpG-rich promotor regions of ubiquitously expressed (housekeeping) genes. In tissue-specific genes lacking CpG clusters (although lack of

CpG islands is not a unique feature of tissue-specific genes [10]), a correlation between DNA hypermethylation of gene regions and the inactive state was not detected in the human factor IX gene [336] or could be demonstrated for only a single methylation site in the mouse ornithine carbamoyl transferase gene [337]. Therefore, the dosage compensation of tissue-specific genes may be achieved by other control elements and not by DNA methylation.

The process of DNA methylation lags behind the inactivation event [338]. HPRT sequences of the first intron of this gene were found to remain unmethylated in postimplantation mouse embryos several days after X chromosome inactivation had occurred. Although these results do not exclude that methylation of sequences, other than the examined ones, may play a role in those initial phases of the inactivation process, they strengthen the hypothesis that DNA methylation serves as a security mechanism helping to maintain the inactive state. According to these results, it is likely that during the initial inactivation process other events not related to de novo DNA methylation are involved [338]. The nature of these events and also the molecular basis of the spreading of X chromosome inactivation are completely unknown. It may be speculated that the interaction of X chromosome-specific DNA binding proteins with regulatory sequences on the X chromosome are involved in these processes. The experiments of Goldman et al. [339] using transgenic mice with X chromosome insertions of a chicken transferrin gene have shown that such sequences must exist and that the inactivation event cannot spread over a distance of 187 kilo base pairs in the absence of a specific signal sequence. Binding and presence of these postulated proteins may lead to a de novo methylation of these sequences in the sense of a fail-safe mechanism. A two-step model has been postulated for mammalian X chromosome inactivation. It involves heterochromatization and DNA methylation as the two inactivating mechanisms [340]. Both mechanisms can function per se alone. This is shown by the gene transfer experiments using naked DNA and by the absence of methylation as an inactivating imprint in extraembryonic tissues [244]. It should be noted that in extraembryonic tissues of eutherian mammals and in marsupial mammals, the paternal X chromosome is not stably repressed and it was shown that X chromosome inactivation in these tissues does not involve DNA hypermethylation [244, 326]. Thus, it is reasonable to think that DNA methylation acts as a stabilizing or locking mechanism in X chromosome inactivation. Although the inactivation is quite stably inherited and spontaneous reactivations are rare in tissues of the adult [341], it was recently demonstrated that the frequency of reactivation events increases significantly with aging. Wareham et al. [342] have used mice carrying X autosomal translocations, where the normal X chromosome is non-randomly inactivated and the translocated chromosome carries additionally an X-linked mutant ornithine carbamoyl transferase (OCT) gene coding for an enzyme which is inactive at pH 7. They showed by histochemical techniques that the number of enzyme positive liver cells (i.e. the cells with reactivated wild type OCT loci derived from the normal X chromosome) increased parallel to the aging of mice. The frequency of X chromosome reactivation was shown also to increase after treatment of cells with 5-azacytidine [reviewed in [343]]. On the other hand, DNA methylation decreases with age in fibroblasts and living mice (see 7.14). Thus, the age-related reactivation of an X-linked gene may be related to DNA demethylation events.

7.9 Genomic Imprinting

At fertilization, the diploid organisms is created from two different haploid sets of chromosomes. Paternally and maternally derived chromosomes cooperate in the development of the new organism. In mice, both the maternal and paternal genome is essential for normal development [344–347]. Haploid parthenogenetic eggs which receive a male pronucleus develop normally but those with two female pronuclei show incomplete development after implanitation [346]. Throughout development, a memory of paternal or maternal origin of each chromosome exists. This imprinting may be related to the failure of parthenogenesis in mammals. The preferential inactivation of the paternally derived X chromosome in extraembryonic lineages of rodents and in marsupial mammals [326] are other examples of genomic imprinting. The imprinting mechanism, whatever its molecular basis might be, should have the following characteristics: It must be heritable through DNA replication; it must be possible to erase the imprinted state prior to gametogenesis in order to rebuild a new differential imprint in male and female gametes. Incomplete erasure and rebuilding would result in transmission of a grandparental component to the next generation. DNA methylation appears to be a likely candidate for this imprinting process, and it was demonstrated recently that it is actually involved. The methylation state of exogenous DNA sequences in transgenic mice was shown to vary according to maternal or paternal origin of the gamete [348–350]. In addition, expression of a transgenic c-myc construct was dependent on its methylation state concomitantly with the parental legacy of the gamete [350]. Methylation of the transgene was observed after passage through a female parent and the methylated state was lost after gametogenesis in the male. The transgene was expressed only in the heart of male mice and was undermethylated [350]. In other cases, the transgene was also less methylated, when derived from the father [348, 349]. Accordingly, in a transgenic mouse strain containing the hepatitis B surface antigen gene, the expression of this gene was inhibited after passage through the female germ line accompanied by a de novo methylation process. However, in contrast to the above mentioned switching of the methylation pattern from its male to female "configuration" in successive generations, the de novo methylation of the hepatitis B surface antigen transgene in the female germ line was irreversible [351]. Methylation patterns of a transgene may be influenced by the chromatin configuration of the surrounding chromosomal regions, a position effect influencing methylation and expression [9]. The methylation-connected imprinting mechanism operating on transgenes is likely to operate also at endogenous loci. Yet, with the exception of X-linked genes, which are methylated to a different extent in some regions of the gene on active and inactive X chromosomes (see 7.8), no allel-specific methylation differences have been found due to difficulties in distinguishing between alleles on southern blots. By analysis of a naturally occurring restriction fragment length polymorphism (RFLP) in the variable tandem repeat of the human c-Ha-ras-1 gene, Chandler et al. [352] detected for the first time methylation differences at allelic loci in fetal fibroblasts and immortal cell lines. This may encourage the search for other allel-specific methylation patterns in the human genome, since it is known that some human diseases associated with autosomes appear to be inherited preferentially from male or female parents [350].

7.10 Agents that Affect DNA Methylation

The DNA methylation process can be modified experimentally by a variety of effectors. A modulation of the metabolism of AdoMet and *S*-adenosylhomocysteine (AdoHcy) is expected to cause changes in the DNA methylation level. In fact, AdoHcy is an inhibitor of purified DNA methyltransferase [157] and it inhibits the methylation process also in permeabilized cells [92]. Thus, modulations of cellular DNA methylation processes may be a consequence of an altered AdoMet/AdoHcy ratio. Methylthioadenosine, a catabolite of AdoMet, decreases DNA methylation levels in transformed rat kidney cells, yet increases the ratio of AdoMet/AdoHcy [353]. Treatment of cells with periodate-oxidized adenosine, an inhibitor of AdoHcy hydrolase, results in a decrease of DNA cytosine methylation [354]. There are in fact only few reports showing a clear direct correlation between altered AdoMet/AdoHcy ratios and altered DNA methylation levels [355, 356]. Evidence was obtained that DNA hypomethylation, resulting after feeding of rats with a methyl-deficient carcinogenic diet, may be linked to decreased AdoMet levels [357]. The relationship between altered methionine metabolism and transmethylation reactions in cancer has been extensively discussed [358, 359]. Other potent inhibitors of DNA methylation are *S*-adenosylethionine, an analogue of AdoMet, and its precursor ethionine, which is a known carcinogen [167, 360–362].

The most extensively studied DNA methylation inhibiting agent is 5-azacytidine. The biological effects of this drug are discussed in Chapts. 7.5, 7.7, 7.8, 7.11, 7.13, 7.14.

It was already mentioned that many carcinogens are DNA hypomethylating agents. The experimental evidence will be discussed below (7.13) and has been summarized earlier [358, 359, 363, 364].

Only a few agents are known to induce DNA hypermethylation. Some DNA synthesis inhibitors are known to provoke hypermethylation of DNA up to 150% of the control levels at concentrations leading to inhibition of DNA synthesis. Among them are aphidicolin [365], cytosine arabinoside [366–368], and hydroxyurea [365, 368, 369]. Assuming that the concentration of DNA methyltransferase is limited within the cell [189], it is possible that a sudden block of the replication process may allow the excess of DNA methyltransferase molecules, which are no longer needed for maintenance methylation, to methylate additional sites de novo. Sodium butyrate, another inhibitor of cell growth, causes an increase of the DNA methylation level to more than 200% of the control levels in human embryonic lung fibroblasts [370]. Furthermore, the enzymatic methylation of DNA, modified by 4-acetoxyamino-quinoline-1-oxide [371] and by *N*-2-aminofluorene [372], is increased. Another interesting example of experimentally induced DNA hypermethylation is observed after ethylation of DNA polymers with ethyl methanesulfonate. The presence of the ethyl group at *N*-7 of guanine or at a dioxyphosphate oxygen seems to mimic a hemi-methylated site and results in increased de novo methylation [373].

7.11 Cell Differentiation and DNA Methylation

Differential DNA methylation patterns are frequently found in particular regions of expressed and non-expressed genes (see 7.7). Therefore, it is highly probable that changes of DNA methylation are involved in processes of cell differentiation and development. Cell differentiation can be artificially induced in a number of systems by treatment of cells with the cytidine-analogue 5-azacytidine or its 2′-deoxyderivate. It was initially found that 5-azacytidine induces the formation of contractile striated muscle cells in mouse 10T1/2 cells [374]. Later it was shown that formation of differentiated adipocytes and chondrocytes was also a consequence of treatment of mouse embryo cells [316, 375] or hamster embryo cells [376] with 5-azacytidine. The cytidine-analogue induces differentiation phenomena also in transformed cells. For example, maturation of the human promyelocytic leukemia cell line HL60 [377, 378] and the Friend erythroleukemia cell line [379] was observed after treatment with 5-azacytidine. Morphological and functional differentiation was also induced in cells of acute myeloid leukemias following exposure to 5-aza-2′-deoxycytidine [380]. In contrast, treatment of the leukemic cell line K562 with 5-azacytidine does not alter the state of differentiation in these cells [381]. Finally, the differentiation of a human melanoma cell line in presence of 5-azacytidine was accompanied by DNA hypomethylation [368]. In some cases, the phenotypical changes induced by 5-azacytidine seem to be not just restricted to the developmental lineages the treated cells are derived from. In several mesenchymal cell lines, 5-azacytidine is able to induce the formation of proteins normally specific for other tissues. Cytokeratin-containing cells as well as myoblasts and astrocytes were obtained from a teratocarcinoma-derived pre-adipocyte cell line [382].

Some other DNA hypomethylating agents are also known to provoke differentiation processes. Among them are ethionine, which induces differentiation in erythroleukemia cells [360], as well as pseudoisocytidine and 5-fluoro-2′-deoxycytidine, which show similar effects like 5-azacytidine in 10T1/2 cells [366]. 3-deaza-(+)-aristeromycin, an inhibitor of AdoHcy hydrolase, is effective in HL60 cells [383], sodium selenite [384] causes differentiation in mouse Friend cells and methinin, a naturally occuring methylation inhibitor, induces the production of hemoglobin in K562 cells [385]. Another example is the differentiation of a myoblast cell line in presence of 3-deazaadenosine, a potent inhibitor of AdoHcy hydrolase [386].

The changes of DNA methylation in the course of differentiation processes have been most frequently studied in mouse teratocarcinoma and erythroleukemia cell lines. A differentiation-linked DNA hypomethylation was initially described in teratocarcinoma cells by Fabricant et al. [387]. Later other investigators found that upon treatment with retinoic acid, which induces the formation of parietal or visceral endoderm in mouse F9 teratocarcinoma cells, a genome-wide demethylation accompanied differentiation [192, 288, 388]. It is of interest to note that 5-azacytidine and retinoic acid treatment of these cells resulted in hypomethylation of similar methylation sites. However, the 5-azacytidine-induced hypomethylation was not accompanied by morphological changes [288]. These data are consistent with another report describing the inability of 5-azacytidine to activate the Moloney murine leukemia virus genome in undifferentiated murine teratocarcinoma cells as opposed to differentiated cells [389]. Taken together, these results indicate that demethylation alone is obviously not sufficient to cause differentiation in teratocarcinoma cells.

Differentiation in erythroleukemia cells can be induced by a variety of chemically unrelated compounds [390]. When the levels of DNA methylation in uninduced cells and in cells induced to differentiate in presence of hexamethylene-bis-acetamide or dimethyl sulfoxide were compared, no difference was observed [391–394]. There was, however, a massive hypomethylation in differentiated cells obtained by treatment with inhibitors of DNA methylation [360, 379]. A closer examination of the DNA methylation levels in the course of Friend cell differentiation has revealed that DNA hypomethylation is a general phenomenon occurring in the presence of any differentiation inducer [395]. This hypomethylation is of transient nature and precedes in all cases the appearance of hemoglobin. Later it was shown that the initial demethylation event is an active process occurring in the absence of DNA replication, whereby 5-methylcytosine is replaced by cytosine [396].

Changes of DNA methylation have also been observed in differentiating melanoma cells. Here, the different inducers exert individual effects on the DNA methylation levels, which are generally not related [368, 397].

The information describing total DNA methylation levels and occurence of DNA methylation within specific genes in the developing embryo is naturally scarce. No global differences between methylation levels in mouse embryo DNA and DNA from adult tissues were found [398], with the exception of extraembryonic tissues, the DNA of which contains about 30% less 5-methylcytosine than the early-cleaving embryo [190–193]. DNA methylation levels at 5′-CCGG-3 sites have been analyzed by a sensitive technique in developing mouse embryos [176]. Sperm DNA was found to be relatively hypermethylated as compared to egg DNA. A loss of genomic methylation was observed during preimplantation development [176], although preimplantation embryos contain a high level of de novo DNA methyltransferase activity [399]. Embryonic and extraembryonic lineages become progressively and independently methylated and the DNA of fetal germ cells stays relatively undermethylated during development [176]. These massive changes of DNA methylation patterns may be involved in the regulation of the developmental processes.

7.12 Active DNA Demethylation

Analysis of tissue-specific genes in various organisms revealed that these genes are often undermethylated in the tissue of expression while being heavily methylated in sperm DNA [252, 254, 400, 401]. This fact suggests that specific sequences undergo demethylation during tissue development. Often differentiation takes place in absence of cell proliferation or even in absence of DNA replication. Therefore one has to assume that besides the passive form of demethylating DNA by simply inhibiting the maintenance of methyl groups during multiple rounds of replication, an active mechanism has to exist which demethylates DNA in absence of DNA replication. Indeed, such an enzymatic activity was found in nuclear extracts of Friend erythroleukemia cells [402]. However, the authors failed to demonstrate whether this enzyme is a glycosylase or part of an excision repair mechanism. The simple chemical removal of the methyl group from the methylcytosine molecule would require a reductive cleavage of a C—C bond and seems energetically unlikely [403]. Further evidence for an active demethylation mechanism came from a set of experiments involving genomic sequencing which revealed an active demethylation process occurring within the regulatory region of the estradiol-inducible avian vitellogenin II gene [247]. Three CpG sites lying within the estradiol-receptor binding site were initially fully methylated and become demethylated in liver cells of white leghorn roosters shortly after estradiol treatment. This demethylation occurred initially in one strand only and correlated well with the induction of vitellogenin mRNA synthesis. The demethylation of the complementary DNA strand lagged 24 hours behind. All critical CpG sites remained demethylated even after cessation of gene transcription. Also Sullivan & Grainger [404)] observed an active DNA demethylation in postmitotic lens cells during chicken development. A mechanism that may be responsible for active DNA demethylation was recently elucidated [396)]. In these experiments, Friend erythroleukemia cells were induced to terminal differentiation with hexamethylenebisacetamide and the DNA was labeled in vivo with a density label, 5-bromodeoxyuridine, and a radioactive label, deoxy[5-^{3}H]cytidine. Newly replicated DNA (heavy-light) was separated from parental DNA (light-light) by isopycnic centrifugation. Incorporation of deoxy[5-^{3}H]cytidine into parental light-light DNA was found only in induced cells concomitantly with the demethylation of DNA. The incorporation of nucleotides into light-light DNA was restricted to deoxycytidine. The labeling of light-light DNA with deoxy[5-^{3}H]cytidine was transient, as the ^{3}H-label was removed during the period of de novo methylation that followed the demethylation. These results strongly favor the hypothesis that demethylation of the DNA during differentiation is achieved by enzymatic mechanisms, whereby 5-methylcytosine is replaced by cytosine probably by an excision repair mechanism. The enzymes responsible for this process have not yet been identified.

In conclusion, DNA methylation patterns, although clonally stably inherited in proliferating cells, can be extensively altered during cell differentiation, presumably to create a new tissue-specific methylation pattern which may guarantee the regulated expression of differentiation-specific genes. Active DNA demethylation may erase certain methyl groups from specific sites of genes that were formerly repressed and become active during cell differentiation.

7.13 DNA Methylation and Cancer

It is widely accepted that the development of cancer is a multistep event [405)] due to an abnormal expression and/or suppression of "critical" genes [406-409)]. This may either be a consequence of acquired genetic alterations like mutations, amplifications, insertions, deletions or translocations of genes causing quantitatively or qualitatively aberrant gene products [406-409)], or may be due to epigenetic processes like alterations in DNA methylation [11)]. Most types of cancer exhibit a considerable genetic lability that leads to the rapid emergence of tumor cell variants [410, 411)]. Some of these have a selective growth advantage which is manifested by increasingly aggressive or malignant behavior, and will finally lead to metastatic spread [412)]. The rapid emergence of new tumor cell variants may be due to an aberrant expression or suppression of distinct genes caused by either genetic or epigenetic phenomenons, rather than being a consequence of the development of new genes, arising by mutation [11, 413-417)].

Several lines of evidence exist that DNA methylation, at least in part, may be involved in the development, establishment and even progression of cancer: a) chemical carcinogens frequently alter the overall methylation of DNA during carcinogenesis, b) a large number of tumor tissues and malignant cell lines have developed aberrant DNA methylation patterns, c) viral and cellular oncogenes, known to contribute to tumorigenesis, have been found to be aberrantly methylated and expressed in certain malignancies, d) the tumorigenicity of several oncogenic viruses can be suppressed by in vitro methylation of the viral DNA prior to infection, and e) the drug 5-azacytidine, known to act preferentially by causing hypomethylation of DNA, is capable of transforming normal cells and can also influence the biological behavior and metastatic spread of certain tumors.

A number of chemical carcinogens has been described that can affect DNA methylation directly in vitro and in vivo. Initial experiments by Drahovsky and Morris [418)] have shown that chemical modifications of *E. coli* DNA by either depurinisation with formic acid, or deamination of deoxycytidylate residues with sodium bisulfite, or alkylation of deoxyguanylate residues with dimethylsulfate, largely decrease or even abolish the ability of this DNA to accept methyl groups in vitro, enzymatically transferred by rat liver DNA methyltransferase. Likewise, the arylating compounds 2-(acetylamino)fluorene, *N*-acetoxy-*N*-2-acetylaminofluorene (NAAF) and its 7-iodo derivative (NAAIF), highly effective liver carcinogens, bind to DNA, predominantly at position 8 of guanine residues. This causes a decreased methyl group acceptance of this modified DNA which correlates well with the amount of DNA-bound carcinogen [419, 420)]. Similarly, an alkylation of DNA caused by a variety of chemicals, including methylnitrosourea (MNU), *N*-methyl-*N*-nitro-*N*-nitrosoguanidine (MNNG), ethyl methanesulfonate (EMS), ethylnitrosourea (ENU), *N*-ethyl-*N*′-nitro-*N*-nitrosoguanidine (ENNG), 1,3-bis(2-chloroethyl)-1-nitrosourea (BSNU), 9-aminoacridine and nitrogen mustard (HN2), decreases the in vitro substrate efficiency of modified DNAs in a dose-dependent manner [421)]. Also other carcinogens like benzo(a)pyrene diolepoxide, which is the active metabolite of benzo(a)pyrene, cause an inhibition of DNA methylation in vitro [421–424)]. Both in vitro and in vivo [421, 425–428)], the above-mentioned carcinogens cause a strong reduction of DNA methylation of treated cells that remains clonally stable. The kind of interference of alkylating and arylating agents with DNA methylation may involve modification of the DNA substrate as well as a direct inhibition of the enzyme [421, 429, 430)]. While the carcinogen-induced conformational changes of DNA are very heterogenous, including adduct formation, creation of apurinic sites and single-strand breaks, the interaction with DNA methyltransferase molecules may either inhibit the activity of DNA methyltransferase directly by interacting with the sulfhydryl-groups of the enzyme [429,430)], or by preventing the "walking mechanism" [420)], which is the preferred mode of action of eukaryotic DNA methyltransferases [163)]. In vitro studies with hemimethylated and unmethylated DNA have revealed that both maintenance [421, 422)] and de novo activity [418–424, 429, 430)] are equally inhibited by the majority of alkylating agents.

Further evidence that carcinogenesis and aberrant DNA methylation may be linked, derives from studies with the liver carcinogen L-ethionine, an analogue of the amino acid L-methionine. This latter amino acid is the biosynthetic precursor of AdoMet, which serves as the methyl group donor for multiple cellular methylation reactions, including methylation of DNA. In vivo application of L-ethionine causes an intracellular depletion of the L-methionine pool, which results in a major disturbance of all L-methionine-dependent methylation reactions of the cell [358, 359)]. L-Ethionine treatment causes undermthylation of DNA in ethionine-treated cells [360,362,431], probably due to the formation of *S*-adenosyl-L-ethionine, an inappropriate cofactor for transmethylation reactions, which is able to inhibit DNA methylation effectively.

Interestingly, many naturally occurring tumors and established tumor cell lines exhibit major disturbances in their methionine metabolism leading to imbalances of the AdoMet/AdoHcy pool [for review see [358, 359]]. It is therefore possible that during tumor development these AdoMet/AdoHcy imbalances may influence the formation of an aberrant DNA methylation pattern. Even if the disturbances of the intracellular AdoMet/AdoHcy levels are only transient, stable phenotypic changes may result in individual neoplastic cells because of the heritability of the DNA methylation pattern.

5-Azacytidine, the most potent inhibitor of DNA methylation, can induce oncogenic transformation in rodent cells [432-437] and induces tumors in vivo [438-440]. In other systems, however, 5-azacytidine seems to suppress tumorigenesis [441-443]. In experimental mouse tumors, the metastatic phenotype can be induced by treatment with 5-azacytidine [416]. 5-Azacytidine is a potent teratogen in NMRI mice [444]. Moreover, this drug may cause point mutations in yeast [445] and *Drosophila* [446], but seems to be only weakly mutagenic in gene mutation studies using V79 Chinese hamster cells [447]. On the other hand, 5-azacytidine is an effective antileukemic agent in both humans [448] and mice [449-451]. The antineoplastic action of 5-azacytidine depends on its incorporation into DNA [452], and the potential number of 5-azacytidine-induced hemimethylated sites may be important [453]. The antineoplastic effect of 5-azacytidine may be due to differentiation induction of tumor cells (see 7.11) and/or enhanced expression of major histocompatibility antigens. Indeed, in two T-cell lines established from individuals with T-cell acute lymphocytic leukemia (T-ALL), which did not express HLA-DR antigens, treatment of these T-cells with 5-azacytidine resulted in the induction of DR surface antigen expression, concomitant with the partial demethylation of the DRα DNA sequences [454].

It is reasonable to assume that the effects of 5-azacytidine, which appear to be so drastically different (induction of differentiation, tumorigenesis and suppression of tumorigenesis by the same compound) are reflected by the perturbing effects of the drug on orderly gene expression in a given cell type.

It has been documented in many studies that treatment of cells or whole animals with 5-azacytidine results in massive DNA demethylation. This demethylation is frequently accompanied by reactivation of previously silent genes (see 7.7) In one study it was shown that Ha-ras and Ki-ras oncogenes in Chinese embryo fibroblasts undergo extensive demethylation during 5-azacytidine-induced oncogenic transformation [433]. Therefore, the above-mentioned tumorigenic capacity of 5-azacytidine may be, at least in part, a consequence of 5-azacytidine-induced aberrant expression of "critical" genes, for example, cellular or viral oncogenes. Samid et al. [455] produced revertants of NIH 3T3 cells transformed by a long terminal repeat-activated Ha-ras proto-oncogene by prolonged α/β interferon treatment. Treatment either in vitro or in vivo with 5-azacytidine resulted in transformation of the persistent revertants, but not of parental NIH 3T3 cells. The retransformants reverted again under interferon treatment, suggesting that DNA methylation is involved in interferon-induced persistent reversion.

As many carcinogens inhibit DNA methylation, multiple studies were initiated that measured the total genomic DNA methylation in transformed cells as well as their normal counterparts. Summarizing the experimental data, it was almost consistently found that naturally occurring malignant tumors and tumor cell lines are hypomethylated in relation to non-malignant tissues [414, 456–458]. However, some other studies failed to detect alterations in the total DNA methylation levels of tumor material derived from fresh pediatric tumor explants [459], non-metastatic prostatic tumors [460], and acute leukemias [461] in comparison to the corresponding normal tissues.

In accordance with the reduced levels of total DNA methylation, many genes are substantially less methylated in cancer cells. For example, patients with either colon or lung cancers had lower methylation levels of the growth hormone genes, globin genes, γ-crystallin genes and α-chorionic gonadotropin genes in the DNA of the cancer tissue compared to the adjacent normal tissue [462,463]. Similar results were obtained for the α-fetoprotein gene in hepatomas compared to normal mouse liver [226, 286, 464], or the ribosomal RNA genes in virus-transformed rat cells [285], and the intracisternal A particle genes in mouse myeloma cells [465]. In contrast, other studies failed to detect hypomethylation of several genes in malignant cells [466, 467]. Thus, the observed hypomethylation of total DNA in malignant cells does not affect all genes equally, but it is completely unknown which mechanisms are responsible for the selective occurrence of DNA hypomethylation. As it is widely accepted that cellular and viral oncogenes are involved in the development and maintenance of cancer, several groups have studied the methylation patterns of these "critical" genes. Sutter & Doerfler [292] investigated adenovirus type 12-transformed cells and found that early genes, which appear to be the "oncogenes" of the virus, are undermethylated and expressed, whereas late genes remain silent and are heavily methylated. Furthermore, the Moloney sarcoma virus carrying the viral *v-mos* oncogene, can effectively transform NIH3T3 cells unless the long terminal repeat sequences or the viral oncogene are methylated in vitro, which decreases the transformation efficiency to more than 70%. Treatment of the transfected cells with 5-azacytidine could essentially re-transform the cells [281]. Also for the Snyder-Theilen feline sarcoma virus, carrying the *v-fes* oncogene, it was reported that this viral oncogene is much more methylated in revertants than in transformed cells, and 5-azacytidine retransforms revertants [468]. In some transformed cells containing the viral *src* sequence, it has been demonstrated that specific sites of the *env-src* region are essentially undermethylated [469–470]. Kuo et al. [471] analyzed a human breast adenocarcinoma cell line, expressing large amounts of RNA sequences related to Kirsten murine leukemia virus. Cell line variants, derived from later passages of transplantation in athymic mice, expressed lower levels of Ki-ras RNA. The expression of the viral RNA sequences in these cell lines correlated inversely with the level of viral DNA methylation in the mouse genome. Feinberg and Vogelstein [472], and Ramsden et al., [473] showed that *ras* oncogenes are hypomethylated in some tumors in comparison to the corresponding normal tissue. In order to investigate whether this hypomethylation of DNA is causatively involved in turmor development or simply an accompanying effect, Borrello et al. [474] methylated in vitro a plasmid containing the transforming Ha-*ras* oncogene by using the bacterial methyltransferases HpaII and HhaI, and used these plasmids for transfection of NIH 3T3 cells. They found that double methylation reduced the transforming activity of the Ha-*ras* containing plasmid by more than 80%. The DNA-demethylating agent 5-azacytidine was able to restore the transforming activity upon 24 h incubation of the transfected NIH 3T3 cells. These data imply that DNA methylation of specific CpG sites can efficiently affect the transforming activity of at least the Ha-*ras* oncogene. Furthermore, Cheah et al.[475] studied the site-specific methylation of the human *c-myc* oncogene in a set of etablished human tumor cell lines and normal human fibroblasts. They found that a region including the first *c-myc* exon, the first intervening sequence, and the second exon was hypomethylated in all cells studied, regardless of being malignant or not, but that inside the third exon a single CCGG site exists which is fully methylated in normal cells and extensively hypomethylated in 3 out of 5 tumor cell lines investigated. As most of these studies were performed by use of methylsensitive restriction endonucleases, only a small subset of methylation sites could be examined in this way. Furthermore, it was impossible to distinguish between fully and hemimethylated sites. Hence, there may exist much more sites in or adjacent to coding regions of genes that may be critical for the regulated expression of these genes, but may be undetectable by restriction endonuclease studies.

In order to analyze whether DNA methylation and metastatic spread of tumors is linked, Liteplo and Kerbel[417] studied a series of metastatic variants of the human melanoma cell line MeWo and found a striking hypomethylation of DNA in the highly metastatic cell lines. These results confirmed the data of Gama-Sosa et al.[414], who showed earlier that the level of 5-methylcytosine of DNA isolated from more than 100 different human tumors, is significantly lower in metastatic neoplasms than in benign neoplasms, non-metastatic tumors and normal tissues. Also in other malignancies, DNA demethylation seems to be linked to progression of the disease[460, 461]. On the other hand, studies from several laboratories have shown that exposure of nonmetastatic malignant cells to 5-azacytidine leads to a small subpopulation of malignant cells having acquired the new ability to metastasize when implanted into the appropriate host[416, 476–478]. These results suggest that 5-aza-C-mediated DNA demethylation leads to activation of genes associated with the metastatic phenotype which are repressed in nonmetastatic cells. Yet the nature of these genes is completely unknown.

In contrast, the tumorigenicity of tumor cells can be efficiently decreased by treatment with various mutagens[479] and the weakly or nonmutagenic compound 5-azacytidine[441]. Arising clonal cell populations, which are incapable of forming progressively growing tumors in syngenic animals, are classified as tum$^-$ variants[480]. These variants express new surface antigens, termed tum$^-$ antigens, which cause an increase in immunogenicity of tum$^-$ cells[441, 479]. The newly expressed surface antigens are recognized by cytolytic T lymphocytes but usually fail to induce any detectable antibody response[481]. As multiple alkylating mutagens but also 5-azacytidine substantially inhibit the methylation process of DNA, Frost et al.[441] have postulated that DNA hypomethylation may be an important step in the induction of the tum$^-$ phenotype, probably via transcriptional activation of usually silent genes. Indeed, Altevogt[442] was able to show that *N*-methyl-*N*-nitro-*N*-nitrosoguanidine- or 5-azacytidine treatment of murine lymphoma Eb cells decreases the total genomic DNA methylation level in some but not all tum$^-$ clones analyzed. As newly-arising tum$^-$ antigens confer the increased immunogenicity of tum$^-$ cells, it is of considerable importance to define the molecular nature of these antigens and to assess their degree of methylation. Recently, De Plaen et al.[482] cloned a tum$^-$ antigen (P91A) of P815 mouse mastocytoma cells, and revealed that a point mutation in one of the several exons of the P91A gene was responsible for the expression of the tum$^-$ phenotype. However, no analysis of the DNA methylation pattern of this gene was carried out, and it is unclear as well, whether this finding is representative for all tum$^-$ variants. At least the high frequency of stable tum$^-$ variants, which arise in 1 to 20% of surviving, mutagen-treated tumor cell populations[479], indicates that epigenetic phenomenons, e.g., DNA methylation, may be involved.

7.14 DNA Methylation and Aging

Holliday and Pugh[5] proposed a mechanism by which progressive methylation of a repetitive DNA sequence would act as a biological clock counting cell divisions. A special "clock" methyltransferase may start to modify one complementary pair of cytosines at one end of the repeated sequence at each round of DNA replication. Depending on the number of methylatable sites, methylation will be completed after a fixed number of cell divisions. If those sequences would be part of a control region of a gene, complete methylation would result in the silencing of this gene and may cause cell differentiation or cell death. However, no evidence has been reported supporting this hypothesis.

Scarano[483] has proposed that 5-methylcytosine may be deaminated to thymine in a programmed manner throughout tissue development. Such deamination has been demonstrated to occur during base analysis in vitro as well as in bacteria where mutational hotspots can be related to the presence of 5-methylcytosine[127]. However, such nonenzymatic deamination is a random process rather than a step in a regulated differentiation program. Evidence for similar mutational events occurring in vertebrates during evolution may originate from the frequent appearance of restriction fragment length polymorphisms detected with restriction enzymes which contain CpGs in their recognition sequence[484].

Evidence that DNA methylation may be involved in the programmed senescence of cells and tissues was presented by Wilson and Jones[485], and Shmookler Reis and Goldstein[486], who found that the 5-methylcytosine content in DNA of diploid fibroblasts from mice, hamsters and humans decreased markedly with age in cell culture. The greatest loss of 5-methylcytosines was observed in mouse fibroblasts which had the most limited lifespan. Immortal cell lines had a more stable DNA methylation pattern.

Some data are available on possible changes of DNA methylation during aging in vivo: Romanov and Vanyushin[487] reported significant losses of DNA methylcytosine residues in old aged cows. DNA methylation in short living species of mice did decline more rapidly than in long-living relatives[488]. Ono et al.[489] have found substantial hypomethylation of spleen DNA in aged mice, whereas brain and liver DNA showed no change or even hypermethylation. Ehrlich et al.[122] failed to detect any decrease in DNA methylation of tissues from young and old humans. Also in vitro this group found no significant changes in the 5-methylcytosine content of human fibroblasts propagated over many cell doublings.

Some scientists have investigated age-related changes of DNA methylation of certain genes and found that the vitellogenin gene in roosters [490], the intracisternal A particle gene [465] and the *c-myc* proto-oncogene in mice [489] became hypomethylated during the aging process.

Recent experimental data have shown that inhibition of DNA methylation during cell propagation can accelerate the cell senescence directly. In fact, when 5-azacytidine was used to induce loss of 5-methylcytosines in DNA, the in vitro lifespan of the cells was considerably shortened [491,492].

Summarizing the experimental data available at present, one has to conclude that the rate of loss of DNA 5-methylcytosine residues appears to be inversely related to the lifespan of cells, tissues and animals. However, it is still unknown whether the decline of DNA methylation levels is the cause for or the consequence of senescence.

7.15 Other Functions

In addition to the summarized functions of DNA methylation which are all more or less related to its involvement in regulation of gene expression, other possible roles have been attributed to DNA methylation in eukaryotes. Only a few specific 5-methylcytosines seem actually to be involved in gene suppression. The function of the bulk of 5-methylcytosines is still obscure. Some of the proposals were derived from already known functions of DNA methylation in prokaryotes [see 2]. There is no convincing evidence that site-specific restriction endonucleases exist in cells of higher eukaryotes and thus participation of DNA methylation in restriction modification processes appears highly unlikely.

The direction of mismatch repair processes by DNA methylation is well documented in bacteria. Since also in eukaryotes at least the delayed methylation lags significantly behind DNA synthesis, a possible role of DNA methylation could be also strand discrimination in mismatch repair. Using hemimethylated SV40 DNA constructs containing single-base-pair mismatches, Hare and Taylor have shown that hemimethylation of two sites adjacent to the mismatches was able to direct the repair to occur only on the unmethylated strand [493].

It is also speculated that methylation plays a role in recombination processes. Methylation of satellite DNA is decreased in DNA of spermatocytes [494–497]. Undermethylation of satellite DNA was observed throughout spermatogenesis in the mouse and in oocytes [498,499]. The rearrangement of immunoglobulin genes is accompanied by DNA demethylation [259–261]. Treatment of cells with 5-azacytidine increases the frequency of rearrangements associated with a malignant phenotype [433].

Finally it has been speculated that DNA methylation is somehow involved in replication processes. Hemimethylated sites may inhibit the replication process preventing a reinitiation of DNA synthesis before methylation is complete. In *E. coli*, hemimethylation actually inhibits DNA replication [500] and in case of the P1 plasmid it was shown that the postreplicational methylation of the origin is required for reinitiation of replication [501].

8 Acknowledgements

We would like to express our gratitude to Prof. L. Träger for critically reading this manuscript and stimulating discussions. The original work was supported by grants from the Deutsche Forschungsgemeinschaft (Dr 104/8) and from the Deutsche Krebshilfe, Mildred Scheel Stiftung e. V. (W25/86/Ho 1) to Dušan Drahovsky.

9 Abbreviations

AdoHcy	*S*-adenosyl-L-homocysteine
AdoMet	*S*-adenosyl-L-methionine
ATP	adenosinetriphosphate
HPRT	hypoxanthine phosphoribosyltransferase
HPV	human papilloma virus
HTF	HpaII tiny fragments
tk	thymidine kinase

10 References

1. Hotchkiss RD (1948) J. Biol. Chem. 175: 315
2. Wyatt GR (1951) Biochem. J. 48: 584
3. Srinivasan PR, Borek E (1964) Science 145: 548
4. Riggs AD (1975) Cytogenet. Cell Genet. 14: 9
5. Holliday R, Pugh JE (1975) Science 187: 226
6. Cooper D (1983) Hum. Genet. 64: 315
7. Doerfler W (1983) Annu. Rev. Biochem. 52: 93
8. Razin A, Szyf M (1984) Biochim. Biophys. Acta 782: 331
9. Jaenisch R, Jähner D (1984) Biochim. Biophys. Acta 782: 1
10. Bird AP (1986) Nature 321: 209
11. Holliday R (1987) Science 238: 163
12. Cedar H (1988) Cell 53: 3
13. Taylor JH (1984) DNA methylation and cellular differentiation, Springer, Berlin Heidelberg New York
14. Adams RLP, Burdon RH (1985) Molecular biology of DNA methylation, Springer, Berlin Heidelberg New York
15. Razin A, Cedar H, Riggs AD (1984) DNA methylation: biochemistry and biological significance, Springer, Berlin Heidelberg New York
16. Ganesan AT, Hoch J (1982) Molecular cloning and gene regulation in Bacilli, Academic, New York
17. Korch C, Hagblom P, Normark S (1983) J. Bacteriol. 155: 1324
18. Smith HO, Kelly SV (1984) In: Razin A, Cedar H, Riggs AD (eds) DNA methylation, biochemistry and biological significance, Springer, Berlin Heidelberg New York, p 39
19. Bickle TA (1982) In: Linn S, Roberts RJ, (eds) Nucleases Cold Spring Harbor Laboratory, Cold Spring Harbor, New York, p 85
20. Modrich P, Roberts RJ (1982) In: Linn S, Roberts RJ (eds) Nucleases, Cold Spring Harbor Laboratory, New York, p 109

21. Brooks JE, Blumenthal RM, Gingeras TR (1983) Nucl. Acids Res. 11: 837
22. Janulaitis A, Klimasauskas S, Petrusyte M, Butkus V (1983) FEBS Lett. 161: 131
23. Ehrlich M, Gama-Sosa MA, Carreira LH, Ljungdahl LG, Kuo KC, Gehrke CW (1975) Nucl. Acids Res. 13: 1399
24. Pukkila P, Peterson J, Herman G, Modrich P, Meselson M (1983) Genetics 104: 571
25. Glickman BW, Radman M (1980) Proc. Natl. Acad. Sci. USA 77: 1063
26. Schlagman SL, Hattman S, Marinus MG (1986) J. Bacteriol. 165: 896
27. Dohet C, Wagner R, Radman M (1986) Proc. Natl. Acad. Sci. USA 83: 3395
28. Arber W (1974) In: Cohn WE (ed) Progress in nucleic acid research and molecular biology. Academic, New York, p 1
29. Smith HO (1979) Science 205: 455
30. Lark C, Arber W (1970) J. Mol. Biol. 52: 337
31. Yuan R, Hamilton DL (1984) In: Razin A, Cedar H, Riggs AD (eds) DNA methylation, biochemistry and biological significance. Springer, Berlin Heidelberg New York, p 11
32. Haberman A (1974) J. Mol. Biol. 89: 545
33. Hadi SM, Bächi B, Iida S, Bickle TA (1983) J. Mol. Biol. 165: 19
34. Reiser J, Yuan R (1977) J. Biol. Chem. 252: 451
35. Kauc L, Piekarowicz A (1978) Eur. J. Biochem. 92: 417
36. Roberts RJ (1983) Nucl. Acids Res. 11: r135
37. Nelson M, McClelland M (1987) Nucl. Acids Res. 15, Suppl.: r219
38. Hall RH (1971) The modified nucleotides in nucleic acids, Columbia University Press, New York
39. Kan NC, Lautenberger JA, Edgell MH, Hutchinson III CA (1979) J. Mol. Biol. 130: 191
40. Geier GE, Modrich P (1979) J. Biol. Chem. 254: 1408
41. Schlagman S, Hattman S, May MS, Berger L (1976) J. Bacteriol. 126: 990
42. Radman R, Wagner R (1984) Curr. Top. Microbiol. Immunol. 108: 23
43. Wagner R, Meselson M (1976) Proc. Natl. Acad. Sci. USA 73: 4135
44. Glickman BW, van den Elsen P, Radman M (1978) Mol. Gen. Genet. 163: 307
45. Lark C (1968) J. Mol. Biol. 31: 389
46. Korba BE, Hays JB (1982) Cell 28: 531
47. Marinus MG, Morris NR (1974) J. Mol. Biol. 85: 309
48. Szyf M, Gruenbaum Y, Urieli-Shoval S, Razin A (1982) Nucl. Acids Res. 10: 7247
49. Marinus MG (1984) In: Razin A, Cedar H, Riggs AD (eds) DNA methylation, biochemistry and biological significance. Springer, Berlin Heidelberg New York, p 81
50. Raleigh EA, Wilson G (1986) Proc. Natl. Acad. Sci. USA 83: 9070
51. Wagner EF, Auer B, Schweiger M (1979) J. Virol. 29: 1229
52. Auer B, Schweiger M (1984) J. Virol. 49: 588
53. Scherzer E, Auer B, Schweiger M (1987) J. Biol. Chem. 262: 15225
54. Revel HR (1967) Virology 31: 688
55. Swinton D, Hattman S, Crain PF, Cheng CS, Smith DL, McCloskey JA (1983) Proc. Natl. Acad. Sci. USA 80: 7400
56. Toussaint A (1976) Virology 70: 17
57. Ray DS, Hanawalt PC (1964) J. Mol. Biol. 9: 812
58. Rae PMM (1976) Science 194: 1062
59. Cummings DJ, Tait A, Goddard JM (1974) Biochim. Biophys. Acta 374: 1
60. Rae PMM, Spear BB (1978) Proc. Natl. Acad. Sci. USA 75: 4992
61. Gorovsky MA, Hattman S, Pleger GL (1973) J. Cell. Biol. 56: 697
62. Hattman S, Kenny C, Berger L, Pratt K (1978) J. Bacterial. 135: 1156
63. Blackburn EH, Pan WC, Johnson CC (1983) Nucl. Acids Res. 11: 5131
64. Singer Harrison G, Karrer KM (1985) Nucl. Acids Res. 13: 73
65. Ammerman D, Steinbrück G, Bauer R, Wöhlert H (1981) Eur. J. Cell Biol. 24: 154
66. Antequera F, Tamane M, Villanueva JR, Santos T (1984) J. Biol. Chem. 259: 8033
67. Tamame M, Antequera F, Villanueva JR, Santos T (1983) Mol. Cell. Biol. 3: 2287
68. Proffitt JH, Davie JR, Swinton D, Hattman S (1984) Molec. Cell. Biol. 4: 985
69. Bull JH, Wootton JC (1984) Nature 310: 701
70. Jupe ER, Magill JM, Magill CW (1986) J. Bacteriol. 165: 420

71. van Etten JL, Schuster AM, Girton L, Burbank DE, Swinton D, Hattman SH (1985) Nucl. Acids Res. 13: 3471
72. Shapiro HS (1976) In: Fasman GD (ed) Handbook of biochemistry and molecular biology: nucleic acids. CRC, Cleveland, p 259
73. Wagner I, Capesius I (1981) Biochim. Biophys. Acta 654: 52
74. Bonen L, Huh TY, Gray MW (1980) FEBS Letters 111: 340
75. Deumling B (1981) Proc. Natl. Acad. Sci. USA 78: 338
76. Gruenbaum Y, Naveh-Many T, Cedar H, Razin A (1981) Nature 292: 860
77. Ellis THN, Goldsbrough PB, Castleton JA (1983) Nucl. Acid Res. 11: 3047
78. Gelvin SB, Karcher SJ, DiRita VS (1983) Nucl. Acids Res. 11: 159
79. Nick H, Bowen B, Ferl RJ, Gilbert W (1986) Nature 319: 243
80. Bird AP, Taggart MH (1980) Nucl. Acids Res. 8: 1485
81. Adams RLP, McKay EL, Craig LM, Burdon RH (1979) Biochim. Biophys. Acta 563, 72
82. Rae PMM, Steele RE (1979) Nucl. Acids Res. 6: 2987
83. Gruenbaum Y, Stein R, Cedar H, Razin A (1981) FEBS Lett. 124: 67
84. Achwal CW, Ganguly P, Chandra HS (1984) EMBO J. 3: 263
85. Scarbrough K, Hattman S, Nur U (1984) Molec. Cell. Biol. 4: 599
86. Urieli-Shoval S, Gruenbaum Y, Sedat J, Razin A (1982) FEBS Lett. 146: 148
87. Adams RLP, Fulton J, Kirk D (1982) Biochim. Biophys. Acta 697: 286
88. Burdon RH, Adams RLP (1969) Biochim. Biophys. Acta 174: 322
89. Lutz D, Grahn H, Kröger H (1973) Z. Naturforsch. 28c: 460
90. Geraci D, Eremenko T, Cocchiara R, Granieri A, Scarano E, Volpe P (1974) Biochem. Biophys. Res. Commun. 57: 353
91. Kappler JW (1970) J. Cell. Physiol. 75: 21
92. Gruenbaum Y, Szyf M, Cedar H, Razin A (1983) Proc. Natl. Acad. Sci. USA 80: 4919
93. Adams RLP (1974) Biochim. Biophys. Acta 335: 365
94. Drahovsky D, Wacker A (1975) Naturwissenschaften 62: 189
95. Woodcock DM, Crowther PJ, Hunter DS, Cooper IA (1983) Biochim. Biophys. Acta 741: 38
96. Davis T, Kirk D, Rinaldi A, Burdon RH, Adams RLP (1985) Biochim. Biophys. Res. Commun. 126: 678
97. Bugler B, Bertaux O, Valencia R (1980) J. Cell. Physiol. 104: 149
98. Adams RLP, Hogarth C (1973) Biochim. Biophys. Acta 331: 214
99. Bird AP (1978) J. Mol. Biol. 118: 49
100. Razin A, Cedar H (1977) Proc. Natl. Acad. Sci. USA 74: 2725
101. Harbers K, Karbers B, Spencer JH (1975) Biochem. Biophys. Res. Commun. 66: 738
102. Macaya G, Cortedas J, Bernardi G (1978) Eur. J. Biochem. 84: 179
103. Drahovsky D, Boehm TLJ, Kreis W (1979) Biochim. Biophys. Acta 563: 28
104. Bird AP, Taggart MH, Smith BA (1979) Cell 17: 889
105. Whittaker PA, McLachlan A, Hardman N (1981) Nucl. Acids Res. 9: 801
106. Grippo P, Iaccarino M, Parisi E, Scarano E (1968) J. Mol. Biol. 36: 195
107. Sinsheimer RL (1955) J. Biol. Chem. 215: 569
108. Pech M, Steeck RE, Zachau HG (1979) Cell 18: 883
109. Salomon R, Kaye AM (1970) Biochim. Biophys. Acta 204: 340
110. Woodcock DM, Crowther PJ, Diver WP (1987) Biochim. Biophys. Res. Commun. 145: 888
111. Selker EU, Jensen BC, Richardson GA (1987) Science 238: 48
112. Bolden AH, Nalin CM, Ward CA, Poonian MS, McComas WW, Weissbach A (1985) Nucl. Acids Res. 13: 3479
113. Bolden AH, Nalin CM, Ward CA, Poonian MS, Weissbach A (1986) Mol. Cell. Biol. 6: 1135
114. Boehm TLJ, Drahovsky D (1979) Z. Naturforsch. 34: 558
115. Browne MJ, Cato ACD, Burdon RH (1978) FEBS Lett. 91: 69
116. Schneiderman MH, Billen D (1973) Biochim. Biophys. Acta 308: 352
117. Salomon R, Kaye AM (1969) J. Mol. Biol. 43: 581
118. Miller OJ, Schnedl W, Allen J, Erlanger BF (1974) Nature 251, 636
119. Gautier F, Bunemann H, Grotjahn L (1977) Eur. J. Biochem. 80: 175
120. Manuelidis L, Wu JD (1978) Chromosoma 66: 1

121. Schnedl W, Dev VG, Tantravahi R, Miller DA, Erlanger BF, Miller OJ (1975) Chromosoma 59: 59
122. Ehrlich M, Gama-Sosa MA, Huang LH, Midgett RM, Lo KC, McCune RA, Gehrke C (1982) Nucl. Acids Res. 10: 2709
123. Gama-Sosa MA, Midgett RM, Slagel VA, Gethens S, Kuo KC, Gehrke CW, Ehrlich M (1983) Biochim. Biophys. Acta 740: 212
124. Josse J, Kaiser AA, Kornberg A (1961) J. Biol. Chem. 236: 864
125. Swartz MN, Trautner TA, Kornberg A (1962) J. Biol. Chem. 237: 1961
126. Russell GJ, Walker PMB, Elton RA, Subak-Sharpe JH (1976) J. Mol. Biol. 108: 1
127. Coulondre C, Miller JH, Farabough RJ, Gilbert W (1978) Nature 274: 775
128. Bird AP (1980) Nucl. Acids Res. 8: 1499
129. Brown TC, Jiricny J (1987) Cell 50: 945
130. Adams RLP, Davis T, Rinaldi A, Eason R (1987) Eur. J. Biochem. 165: 107
131. Cooper DN, Taggart MH, Bird AP (1983) Nucl. Acids Res. 11: 647
132. Bird AP, Taggart M, Frommer M, Miller OJ, Macleod D (1985) Cell 40: 91
133. Lavia P, MacLeod D, Bird AP (1987) EMBO J. 6: 2773
134. Keshet I, Yisraeli J, Cedar H (1985) Proc. Natl. Acad. Sci. USA 82: 2560
135. Toniolo D, D'Urso M, Martini G, Persico M, Tufano V, Battistuzzi G, Luzzatto L (1984) EMBO J. 3: 1987
136. Yen PH, Patel P, Chinauld AC, Mohandas T, Shapiro LJ (1984) Proc. Natl. Acad. Sci. USA 81: 1759
137. Wolf SF, Jolly DJ, Lunnen KD, Friedman T, Migeon BR (1984) Proc. Natl. Acad. Sci. USA 81: 2806
138. Nass MMK (1973) J. Mol. Biol. 80: 155
139. Vanyushin BF, Kirnos MD (1974) FEBS Letters 39: 195
140. Dawid IB (1974) Science 184, 80
141. Groot GSP, Kroon AM (1979) Biochim. Biophys. Acta 562: 355
142. Bonen L, Huh TY, Gray MW (1980) FEBS Lett. 111: 340
143. Shmookler Reis RJ, Goldstein S (1983) J. Biol. Chem. 258: 9078
144. Pollack Y, Casir J, Shemer T, Metzger S, Szyf M (1984) Nucl. Acids Res. 12: 4811
145. Benne R, Tabak HF (1986) Trends in Genetics 2: 147
146. Whitfield PR, Spencer D (1968) Biochim. Biophys. Acta 157: 333
147. Baxter R, Kirk JTO (1969) Nature 222: 272
148. Royer HD, Sager R (1979) Proc. Natl. Acad. Sci. USA 76: 5794
149. Burdon RH, Martin BT, Lal BM (1967) J. Mol. Biol. 28: 357
150. Sheid B, Srinivasan PR, Borek E (1968) Biochemistry 7: 280
151. Kalousek F, Morris NR (1968) J. Biol. Chem. 243: 2440
152. Pedrali-Noy G, Weissbach A (1986) J. Biol. Chem. 261: 7600
153. Pfeifer GP, Drahovsky D (1986) FEBS Lett. 207: 75
154. Kalousek F, Morris NR (1969) J. Biol. Chem. 244: 1157
155. Roy PH, Weissbach A (1975) Nucl. Acids Res. 2: 1669
156. Turnbull JF, Adams RLP (1976) Nucl. Acids Res. 3: 677
157. Simon D, Grunert F, v. Acken U, Döring HP, Kröger H (1978) Nucl. Acids Res. 5: 2153
158. Pfeifer GP, Grünwald S, Boehm TLJ, Drahovsky D (1983) Biochim. Biophys. Acta 740: 323
159. Drahovsky D, Morris NR (1971) J. Mol. Biol. 61: 343
160. Pfeifer GP, Grünwald S, Palitti F, Kaul S, Boehm TLJ, Hirth HP, Drahovsky D (1985) J. Biol. Chem. 260: 13787
161. Carotti D, Palitti F, Mastrantonio S, Rispoli M, Strom R, Amato A, Campagnari F, Whitehead EP (1986) Biochim. Biophys. Acta 866: 135
162. Grünwald S, Drahovsky D (1984) Int. J. Biochem. 16: 883
163. Drahovsky D, Morris NR (1971) J. Mol. Biol. 57: 475
164. Bestor TH, Ingram VM (1983) Proc. Natl. Acad. Sci. USA 80: 5559
165. Pfeifer GP, Spiess E, Grünwald S, Boehm TLJ, Drahovsky D (1985) EMBO J. 4: 2879
166. Adams RLP, McKay EL, Craig LM, Burdon RH (1979) Biochim. Biophys. Acta 561, 345
167. Sneider TW, Teague WM, Rogachevsky LM (1975) Nucl. Acids Res. 2: 1685
168. Jones PA, Taylor SM (1981) Nucl. Acids Res. 9: 2933

169. McIntosh LP, Zielinski WS, Kalisch BW, Pfeifer GP, Sprinzl M, Drahovsky D, van de Sande JH, Jovin TM (1985) Biochemistry 24: 4806
170. Santi DV, Norment A, Garrett CE (1984) Proc. Natl. Acad. Sci. USA 81: 6993
171. Wang RYH, Huang LH, Ehrlich M (1984) Nucl. Acids Res. 12: 3473
172. Gruenbaum Y, Cedar H, Razin A (1982) Nature 295: 620
173. Bolden A, Ward C, Siedlecki JA, Weissbach A (1984) J. Biol. Chem. 259: 12437
174. Zucker KE, Riggs AD, Smith SS (1985) J. Cell. Biochem. 29: 337
175. Wigler M, Levy D, Perucho M (1981) Cell 24: 33
176. Monk M, Boubelik M, Lehnert S (1987) Development 99: 371
177. Sano H, Noguchi H, Sager R (1983) Eur. J. Biochem. 135: 181
178. Adams RLP, Gardiner K, Rinaldi A, Bryans M, McGarvey M, Burdon RH (1986) Biochim. Biophys. Acta 868: 9
179. Simon D, Stuhlmann H, Jähner D, Wagner H, Werner E, Jaenisch R (1983) Nature 304: 275
180. Adams RLP, Davis T, Fulton J, Kirk D, Qureshi M, Burdon RH (1984) Current topics in microbiology and immunology vol. 108, Springer, Berlin Heidelberg New York
181. Bestor TH (1987) Nucl. Acids Res. 15: 3835
182. Soslau G, Parker J, Nelson JW (1986) Nucl. Acids Res. 14: 7237
183. Smith SS, Hardy TA, Baker DJ (1987) Nucl. Acids Res. 15: 6899
184. Baker DJ, Hardy TA, Smith SS (1987) Biochem. Biophys. Res. Commun. 146: 596
185. Kautiainen TL, Jones PA (1985) Biochemistry 24: 1193
186. Yoo H, Noshari J, Lapeyre JN (1987) J. Biol. Chem. 262: 8066
187. Bestor TH, Ingram VM (1985) Proc. Natl. Acad. Sci. USA 82: 2674
188. Pfeifer GP, Drahovsky D (1986) Biochim. Biophys. Acta 868: 238
189. Drahovsky D, Pfeifer GP (1988) In: Kahl G, (ed) Architecture of eukarytic genes, VCH, Weinheim, F.R.G., p 432
190. Manes C, Menzel P (1981) Nature 293: 589
191. Chapman V, Forrester L, Sanford J, Hastie N, Rossant J (1984) Nature 307: 284
192. Razin A, Webb C, Szyf M, Yisraeli J, Rosenthal A, Naveh-Many T, Sciaky-Gallili N, Cedar H (1984) Proc. Natl. Acad. Sci. USA 81: 2275
193. Rossant J, Sanford JP, Chapman VM, Andrews GK (1986) Develop. Biol. 117: 567
194. DePaoli-Roach A, Roach PJ, Zucker KE, Smith SS (1986) FEBS Lett. 197: 149
195. Creusot F, Christman JK (1981) Nucl. Acids Res. 9: 5359
196. Burdon RH, Qureshi M, Adams RLP (1985) Biochim. Biophys. Acta 825: 70
197. Drahovsky D, Pfeifer GP, Grünwald S, Hirth HP, Vogel M, Brzoska MJ, Palitti F (1985) In: Cantoni GL, Razin A (eds) Biochemistry and biology of DNA methylation, Alan R. Liss, New York, p 67
198. Noguchi H, v. Reddy FP, Pardee AB (1983) Cell 32: 443
199. Ottiger HP, Hübscher U (1984) Proc. Natl. Acad. Sci. USA 81: 3993
200. Morris NR, Pih KD (1971) Cancer Res. 31: 433
201. Szyf M, Kaplan F, Mann V, Giloh H, Kedar E, Razin A (1985) J. Biol. Chem. 260: 8653
202. Kaul S, Drahovsky D (1984) Protides Biol. Fluids 32: 521
203. Kastan MB, Gowans BJ, Lieberman MW (1982) Cell 30: 509
204. Willis DB, Goorha R, Granoff A (1984) J. Virol. 49: 86
205. Thompson JP, Granoff A, Willis DB (1986) Proc. Natl. Acad. Sci. USA 83: 7688
206. Theiss G, Schleicher R, Schimpff-Weiland G, Follmann H (1987) Eur. J. Biochem. 167: 89
207. Sano H, Sager R (1980) Eur. J. Biochem. 105: 471
208. Naveh-Many T, Cedar H (1981) Proc. Natl. Acad. Sci. USA 78: 4246
209. Davie JR, Saunders CA (1981) J. Biol. Chem. 256: 12574
210. Ball DJ, Gross DS, Garrard WT (1983) Proc. Natl. Acad. Sci. USA 80: 5490
211. Weisbrod S (1982) Nature 297: 289
212. Keshet I, Lieman-Hurwitz J, Cedar H (1986) Cell 44: 535
213. Buschhausen G, Wittig B, Graessmann M, Graessmann A (1987) Proc. Natl. Acad. Sci. USA 84: 1177
214. Huang LH, Wang R, Gama-Sosa MA, Shenoy S, Ehrlich M (1984) Nature 308: 293
215. Wang RYH, Zhang XY, Ehrlich M (1986) Nucl. Acids Res. 14: 1599
216. Harrington MA, Jones PA, Imagawa M, Karin M (1988) Proc. Natl. Acad. Sci. USA 85: 2066

217. Singleton CK, Klysik J, Stirdivant SM, Wells RD (1982) Nature 299: 312
218. Behe M, Felsenfeld G (1981) Proc. Natl. Acad. Sci. USA 78: 1619
219. Azorin F, Nordheim A, Rich A (1983) EMBO J. 2: 649
220. Diekmann S (1987) EMBO J. 6: 4213
221. Sager R, Kitchin R (1975) Science 189: 426
222. Ott MO, Sperling L, Cassio D, Levilliers J, Sala-Trepat J, Weiss MC (1982) Cell 30: 825
223. Vorce RL, Goodman JI (1985) Biochem. Biophys. Res. Commun. 126: 879
224. Sellem CH, Weiss MC, Cassio D (1985) J. Mol. Biol. 181: 363
225. Tratner I, Nahon JL, Sala-Trepat JM, Venetianer A (1987) Mol. Cell. Biol. 7: 1856
226. Andrews GK, Dziadek M, Tamaoki T (1982) J. Biol. Chem. 257: 5148
227. Sakuma K, Cook JR, Smith CL, Chiu JF (1987) Biochem. Biophys. Res. Commun. 143: 447
228. Laverriere JN, Muller M, Buisson N, Tougard C, Tixier-Vidal A, Martial JA, Gourdji D (1986) Endocrinol. 118: 198
229. Lieberman MW, Beach LR, Palmiter RD (1983) Cell 35: 207
230. Fukunaga R, Matsuyama M, Okamura H, Nagata K, Nagata S, Sokawa Y (1986) Nucl. Acids Res. 14: 4421
231. Sneller MC, Gunter KC (1987) J. Immunol. 138: 3505
232. Ringuette MJ, Sobieski DA, Chamow SM, Dean J (1986) Proc. Natl. Acad. Sci. USA 83: 4341
233. Winter BB, Arnold HH (1987) J. Biol. Chem. 262: 13750
234. Thompson MD, Nakhasi HL (1985) Cancer Res. 45: 1291
235. Libert F, Vassart G, Christophe D (1986) Biochem. Biophys. Res. Commun. 134: 1109
236. Wilks AF, Cozens PJ, Mattaj IW, Jost JP (1982) Proc. Natl. Acad. Sci. USA 79: 4252
237. Philispsen JNJ, Gruber M, Geert AB (1985) Biochim. Biophys. Acta 826: 186
238. Stein R, Sciaky-Gallili N, Razin A, Cedar H (1983) Proc. Natl. Acad. Sci. USA 80: 2422
239. Benvenisty N, Mencher D, Menyuhas O, Razin A, Reshef L (1985) Proc. Natl. Acad. Sci. USA 82: 267
240. Burch JBE, Weintraub H (1983) Cell 33: 65
241. Gerber-Huber S, May FEB, Westley BR, Felber BK, Hosbach HA, Andres AC, Ryffel GU (1983) Cell 33: 43
242. Grainger RM, Hazard-Leonards RM, Samaha F, Hougan LM, Lesk MR, Thomsen GH (1983) Nature 306: 88
243. McKeon C, Ohkubo H, Pastan I, de Chrombrugghe B (1982) Cell 29: 203
244. Kratzer PG, Chapman VM, Lambert H, Evans RE, Liskay RM (1983) Cell 33: 37
245. Church GM, Gilbert W (1984) Proc. Natl. Acad. Sci. USA 81: 1991
246. Saluz HP, Jost JP (1986) Gene 42: 151
247. Saluz HP, Jiricny J, Jost JP (1986) Proc. Natl. Acad. Sci. USA 83: 7167
248. Maniatis T, Fritsch EF, Lauer J, Lawn RM (1980) Ann. Rev. Genet. 14: 145
249. Waalwijk C, Flavell RA (1978) Nucl. Acids Res. 5: 4631
250. McGhee JD, Ginder GD (1979) Nature 280: 419
251. Shen CKJ, Maniatis T (1980) Proc. Natl. Acad. Sci. USA 77: 6634
252. van der Ploeg LHT, Flavell RA (1980) Cell 19: 947
253. Weintraub H, Larsen A, Groudine M (1981) Cell 24: 333
254. Haigh LS, Owens BB, Hellewell S, Ingram VM (1982) Proc. Natl. Acad. Sci. USA 79: 5332
255. Mavilio F, Giampaolo A, Carè A, Miglianccio G, Calandrini M, Russo G, Pagliardi GL, Mastroberardino G, Marinucci M, Peschle C (1983) Proc. Natl. Acad. Sci. USA 80: 6907
256. Busslinger M, Hurst J, Flavell RA (1983) Cell 34: 197
257. Murray EJ, Grosveld F (1987) EMBO J. 6: 2329
258. Ley TJ, Chiang YL, Haidaris D, Anagnou NP, Wilson VL, Anderson WF (1984) Proc. Natl. Acad. Sci. USA 81: 6618
259. Rogers J, Wall R (1981) Proc. Natl. Acad. Sci. USA 78: 7497
260. Mather EL, Perry RP (1983) Proc. Natl. Acad. Sci. USA 80: 4689
261. Storb U, Arp B (1983) Proc. Natl. Acad. Sci. USA 80: 6642
262. Ott MO, Sperling L, Weiss MC (1984) Proc. Natl. Acad. Sci. USA 81: 1738
263. Orlofsky A, Chasin LA (1985) Molec. Cell. Biol. 5: 214
264. McKeon C, Pastan I, de Chrombrugge B (1984) Nucl. Acids Res. 12: 3491
265. Fernandez MP, Young MF, Sobel ME (1985) J. Biol. Chem. 260: 2374

266. Cate RL, Chick W, Gilbert W (1983) J. Biol. Chem. 258: 6645
267. Petropoulos CJ, Yaswen P, Panzica M, Fausto N (1985) Nucl. Acids Res. 13: 8105
268. Nelson KJ, Mather EL, Perry RP (1984) Nucl. Acids Res. 12: 1911
269. Hjelle BL, Phillips JA, Seeburg PH (1982) Nucl. Acids Res. 10: 3459
270. Vardimon L, Kressmann A, Cedar H, Maechler M, Doerfler W (1982) Proc. Natl. Acad. Sci. USA 79: 1073
271. Waechter DE, Baserga R (1982) Proc. Natl. Acad. Sci. USA 79: 1106
272. Stein R, Razin A, Cedar H (1982) Proc. Natl. Acad. Sci. USA 79: 3418
273. Wigler M, Sweet R, Sim GK, Wold B, Pellicer A, Lacy E, Maniatis T, Silverstein S, Axel R (1979) Cell 16: 777
274. Pollack Y, Stein R, Razin A, Cedar H (1980) Proc. Natl. Acad. Sci. USA 77: 6463
275. Stein R, Gruenbaum Y, Pollack Y, Razin A, Cedar (1982) Proc. Natl. Acad. Sci. USA 79: 61
276. Christy B, Scangos G (1982) Proc. Natl. Acad. Sci. USA 79: 6299
277. Fradin A, Manley JL, Prives CL (1982) Proc. Natl. Acad. Sci. USA 79: 5142
278. Vardimon L, Kuhlmann I, Doerfler W, Cedar H (1981) Eur. J. Cell Biol. 25: 13
279. Hynes NE, Rahmsdorf U, Kennedy N, Fabiani L, Michalides R, Nusse R, Groner B (1981) Gene 15: 307
280. Harbers K, Jähner D, Jaenisch R (1981) Nature 293: 540
281. McGeady ML, Jhappan C, Ascione R, Vande Woude G (1983) Mol. Cell. Biol. 3: 305
282. Graessmann M, Graessmann A, Wagner H, Werner E, Simon D (1983) Proc. Natl. Acad. Sci. USA 80: 6470
283. Macleod D, Bird AP (1983) Nature 306: 200
284. Colgan V, Elbrecht A, Goldman P, Lazier CB, Deeley R (1982) J. Biol. Chem. 257: 14453
285. Kunnath L, Locker J (1983) EMBO J. 2: 317
286. Vedel M, Gomez-Garcia M, Sala M, Sala-Trepat J (1983) Nucl. Acids Res. 11: 4335
287. White R, Parker M (1983) J. Biol. Chem. 258: 8943
288. Young PR, Tilghman SM (1984) Mol. Cell. Biol. 4: 898
289. Eick D, Fritz HJ, Doerfler W (1983) Anal. Biochem. 135: 165
290. Desrosiers RC, Mulder C, Fleckenstein B (1979) Proc. Natl. Acad. Sci. USA 76: 3839
291. Vardimon L, Neumann R, Kuhlmann I, Sutter D, Doerfler W (1980) Nucl. Acids Res. 8: 2461
292. Sutter O, Doerfler W (1980) Proc. Natl. Acad. Sci. USA 77: 253
293. Deuring R, Doerfler W (1983) Gene 26: 283
294. Kruczek I, Doerfler W (1982) EMBO J. 1: 409
295. Guntaka RV, Gowda S, Wagner H, Simon D (1987) FEBS Lett. 221: 332
296. Groudine M, Eisenman R, Weintraub H (1981) Nature 292: 311
297. Guntaka R, Rao P, Mitsialis S, Katz R (1980) J. Virol. 34: 369
298. Cohen JC (1980) Cell 19: 653
299. Youssoufian H, Hammer SM, Hirsch MS, Mulder C (1982) Proc. Natl. Acad. Sci. USA 79: 2207
300. Niwa O, Sugahara T (1981) Proc. Natl. Acad. Sci. USA 78: 6290
301. Szyf M, Eliasson L, Mann V, Klein G, Razin A (1985) Proc. Natl. Acad. Sci. USA 82: 8090
302. Kanamori A, Ikuta K, Ueda S, Kato S, Kanji H (1987) J. Gen. Virol. 68: 1485
303. Hoffman JW, Steffen D, Gusella J, Tabin C, Bird S, Cowing D, Weinberg RA (1982) J. Virol. 44: 144
304. Gautsch JW, Wilson MC (1983) Nature 301: 32
305. Dressler GR, Rock DL, Fraser NW (1987) J. Gen. Virol. 68: 1761
306. Compere SJ, Palmiter RD (1981) Cell 25: 233
307. Clough DW, Kunkel LM, Davidson RL (1982) Science 216: 70
308. Carache S, Dover G, Smith K, Talbot CC, Moyer M, Boyer S (1983) Proc. Natl. Acad. Sci. USA 80: 4842
309. DeSimone J, Heller P, Hall L, Zuriers D (1982) Proc. Natl. Acad. Sci. USA 79: 4428
310. Ley TJ, DeSimone J, Anagou NP, Keller GH, Humphries RK, Turner PH, Young NS, Heller P, Nienhuis AW (1982) New Engl. J. Med. 307: 1469
311. Ivarie RD, Morris JA (1982) Proc. Natl. Acad. Sci. USA 79: 2967
312. Gasson JC, Ryden T, Bourgeois S (1983) Nature 302: 621
313. Mohandas T, Sparkes RS, Shapiro LJ (1981) Science 211: 393

314. Jones PA, Taylor SM, Mohandas T, Shapiro LJ (1982) Proc. Natl. Acad. Sci. USA 79: 1215
315. Graves JAM (1982) Exp. Cell Res. 141: 99
316. Taylor SM, Jones PA (1979) Cell 17: 771
317. Harris M (1982) Cell 29: 483
318. Jeggo PA, Holliday R (1986) Mol. Cell. Biol. 6: 2944
319. Rösl F, Dürst M, zur Hausen H (1988) EMBO J. 7: 1321
320. Gartler SM, Riggs AD (1983) Annu. Rev. Genet. 17: 155
321. Gartler SM, Andina R, Gant N (1975) Exp. Cell Res. 91: 454
322. Takagi N (1974) Exp. Cell Res. 86: 127
323. West JD, Frels WI, Chapman VM, Papaioannou VE (1977) Cell 12: 873
324. Migeon BR, Wolf SF, Axelman J, Kaslow DC, Schmidt M (1985) Proc. Natl. Acad. Sci. USA 82: 3390
325. Lyon MF (1972) Biol. Rev. 47: 1
326. Kaslow DC, Migeon BR (1987) Proc. Natl. Acad. Sci. USA 84: 6210
327. Liskay RM, Evans RJ (1980) Proc. Natl. Acad. Sci. USA 77: 4895
328. Venolia L, Gartler SM, Wassman ER, Yen P, Mohandas T, Shapiro LJ (1982) Proc. Natl. Acad. Sci. USA 79: 2352
329. Wolf SF, Migeon BR (1982) Nature 295: 667
330. Lindsay S, Monk M, Holliday R, Huschtscha L, Davies KE, Riggs AD, Flavell RA (1985) Ann. Hum. Genet. 49: 115
331. Toniolo D, Martini G, Migeon BR, Dono R (1988) EMBO J. 7: 401
332. Lock LF, Melton DW, Caskey CT, Martin GR (1986) Mol. Cell. Biol. 6: 914
333. Riggs AD, Singer-Sam J, Keith DH (1985) In: Cantoni GL, Razin A (eds) Biochemistry and biology of DNA methylation. Alan R. Liss, New York, p 211
334. Keith DH, Singer-Sam J, Riggs AD (1986) Mol. Cell. Biol. 6: 4122
335. Wolf SF, Migeon BR (1985) Nature 314: 467
336. Ruta Cullen C, Hubberman P, Kaslow DC, Migeon BR (1986) EMBO J. 5: 2223
337. Mullins LJ, Veres G, Caskey CT, Chapman V (1987) Mol. Cell. Biol. 7: 3916
338. Lock LF, Takagi N, Martin GR (1987) Cell 48: 39
339. Goldman MA, Stokes KR, Idzerda RI, McKnight GS, Hammer RE, Brinster RL, Gartler SM (1987) Science 236: 593
340. Gartler SM, Dyer KA, Marshall Graves JA, Rocchi M (1985) In: Cantoni GL, Razin E (eds) Biochemistry and biology of DNA methylation. Alan R. Liss, New York, p 223
341. Migeon BR, Wolf SF, Mareni C, Axelman J (1982) Cell 29: 595–600
342. Wareham KA, Lyon MF, Glenister PH, Williams ED (1987) Nature 327: 725
343. Jones PA (1985) Pharmac. Ther. 28: 17
344. Cattanach BM, Kirk M (1985) Nature 315: 496
345. McGrath J, Solter D (1984) Cell 37: 179
346. Surani MAH, Barton SC, Norris ML (1984) Nature 308: 548
347. Surani MAH, Barton SC, Norris ML (1987) Nature 326: 395
348. Sapienza C, Peterson AC, Rossant J, Balling R (1987) Nature 328: 251
349. Reik W, Collick A, Norris ML, Barton SC, Surani MA (1987) Nature 328: 248
350. Swain JL, Stewart TA, Leder P (1987) Cell 50: 719
351. Hadchouel M, Farza H, Simon D, Tiollais P, Pourcel C (1987) Nature: 329: 454
352. Chandler LA, Ghazi H, Jones PA, Boukamp P, Fusenig NE (1987) Cell 50: 711
353. Dante R, Arnaud M, Niveleau A (1983) Biochem. Biophys. Res. Commun. 114: 214
354. Liteplo RG, Kerbel RS (1986) Cancer Res. 46: 577
355. Hershfield MS, Kurtzberg J, Chaffee S, Greenberg ML (1985) In: Cantoni GL, Razin A (eds) Biochemistry and biology of DNA methylation. Liss, New York, p 305
356. Liteplo RG (1988) Cancer Lett. 39: 319
357. Wilson MJ, Shivapurkar N, Poirier LA (1984) Biochem J. 218: 987
358. Hoffman RM (1984) Biochim. Biophys. Acta 738: 49
359. Hoffman RM (1985) Anticancer Res. 5: 1
360. Christman JK, Price P, Pedrinan L, Acs G (1977) Eur. J. Biochem. 81: 53
361. Cox R, Irving CC (1977) Cancer Res. 37: 222
362. Boehm TLJ, Drahovsky D (1979) Eur. J. Cancer 15: 1167
363. Boehm TLJ, Drahovsky D (1983) J. Natl. Cancer Inst. 71: 429

364. Riggs AD, Jones PA (1983) Adv. Cancer Res. 40: 1
365. Nyce J, Liu L, Jones PA (1986) Nucl. Acids Res. 14: 4353
366. Jones PA, Taylor SM (1980) Cell 20: 85
367. Boehm TLJ, Drahovsky D (1982) Cancer Res. 42: 1537
368. Steigerwald SD, Pfeifer GP (1988) Exp. Cell Res.: in press
369. Alvarez E, Elliott BE, Houghton AN, Kerbel RS (1988) Cancer Res. 48: 2440
370. Parker MI, deHaan JB, Gevers W (1986) J. Biol. Chem. 261: 2786
371. Pfohl-Leszkowicz A, Galiegue-Zouitina S, Bailleul B, Loucheux-Lefebvre MH, Dirheimer G (1983) FEBS Lett. 163: 85
372. Pfohl-Leszkovicz A, Fuchs RPP, Dirkheimer G (1984) FEBS Lett. 178: 59
373. Farrance IK, Ivarie R (1985) Proc. Natl. Acad. Sci. USA 82: 1045
374. Constantinides PG, Jones PA, Gevers W (1977) Nature 267: 364
375. Konieczny SF, Emerson jr. CP (1984) Cell 38: 791
376. Sager R, Kovac P (1982) Proc. Natl. Acad. Sci. USA 79: 480
377. Christman JK, Mendelsohn N, Herzog D, Schneiderman N (1983) Cancer Res. 43: 763
378. Momparler RL, Bouchard J, Samson J (1985) Leukemia Res. 9: 1361
379. Creusot F, Acs G, Christman JK (1982) J. Biol. Chem. 257: 2041
380. Pinto A, Attadia V, Fusco A, Ferrara F, Spada OA, DiFiore PP (1984) Blood 64: 922
381. Chen E, Karr RW, Frost JP, Gonwa TA, Ginder GD (1986) Mol. Cell. Biol. 6: 1698
382. Darmon M, Nicolas JF, Lamblin D (1984) EMBO J. 3: 961
383. Aarbakke J, Miura GA, Prytz PS, Bessesen A, Slordal L, Gordon RK, Chiang PK (1986) Cancer Res. 46: 5469
384. Cox R, Goorha S (1986) Carcinogenesis 7: 2015
385. Brockett Lyon S, Buonocore L, Miller M (1987) Mol. Cell. Biol. 7: 1759
386. Scarpa S, Strom R, Bozzi A, Aksamit RR, Backlund Jr. PS, Chen J, Cantoni GL (1984) Proc. Natl. Acad. Sci. USA 81: 3064
387. Fabricant JD, Wagner EF, Auer B, Schweiger M (1979) Exp. Cell Res. 124: 25
388. Bestor TH, Hellewell SB, Ingram VM (1984) Mol. Cell. Biol. 4: 1800
389. Niwa O, Yokota Y, Ishida H, Sugahara T (1983) Cell 32: 1105
390. Marks PA, Rifkind RA (1978) Annu. Rev. Biochem. 47: 419
391. Baur R, Kluge N, Kröger H, Wohlert H (1982) Z. Naturforsch. 37: 812
392. Smith SS, Yu JC, Chen CW (1982) Nucl. Acids Res. 10: 4305
393. Sheffery M, Rifkind RA, Marks PA (1982) Proc. Natl. Acad. Sci. USA 79: 1180
394. Tolberg ME, Smith SS (1984) FEBS Lett. 176: 250
395. Razin A, Feldmesser E, Kafri T, Szyf M (1985) In: Cantoni GL, Razin A (eds) Biochemistry and biology of DNA methylation. Alan R. Liss, New York, p 239
396. Razin A, Szyf M, Kafri T, Roll M, Giloh H, Scarpa S, Carotti D, Cantoni GL (1986) Proc. Natl. Acad. Sci. USA 83: 2827
397. Drahovsky D, Brzoska MJ, Werle A, Boehm TLJ (1985) In: Cantoni GL, Razin A (eds) Biochemistry and biology of DNA methylation. Alan R. Liss, New York, p 271
398. Singer J, Roberts-Ems J, Luthardt FW, Riggs AD (1979) Nucl. Acids Res. 7: 2369
399. Jähner D, Stuhlmann H, Stewart CL, Harbers K, Löhler J, Simon I, Jaenisch R (1982) Nature 298: 623
400. Mandel JL, Chambon P (1979) Nucl. Acids Res. 7: 2081
401. Jones RE, Defeo D, Piatigorsky J (1981) J. Biol. Chem. 256: 8172
402. Gjerset RA, Martin DW jr. (1982) J. Biol. Chem. 257: 8581
403. Cantoni GL (1960) Comp. Biochem. 1: 172
404. Sullivan CH, Grainger RM (1986) Proc. Natl. Acad. Sci. USA 83: 329
405. Foulds L (1975) Neoplastic Development, Academic, New York
406. Klein G, Klein E (1985) Nature 315: 190
407. Knudson AG (1985) Cancer Res. 45: 1437
408. Duesberg PH (1987) Proc. Natl. Acad. Sci. USA 84: 2117
409. Boehm TLJ (1985) In: Progress in clinical biochemistry and medicine 2, Springer, Berlin Heidelberg New York p 1
410. Nowell PC (1976) Science 194: 23
411. Nowell PC (1986) Cancer Res. 46: 2203
412. Poste G, Fidler IJ (1980) Nature 283: 139

413. Fidler IJ, Kripke ML (1977) Science 197: 893
414. Gama-Sosa MA, Slagel VA, Trewyn RW, Oxenhandler R, Kuo KC, Gehrke CW, Ehrlich M (1983) Nucl. Acids Res. 11: 6883
415. Frost P, Kerbel RS (1983) Cancer Metas. Rev. 2: 375
416. Olsson L, Forchhammer J (1984) Proc. Natl. Acad. Sci. USA 81: 3389
417. Liteplo RG, Kerbel RS (1987) Cancer Res. 47: 2264
418. Drahovsky D, Morris NR (1972) Biochim. Biophys. Acta 277: 245
419. Salas CE, Pfohl-Leskowicz A, Lang MC, Dirheimer G (1979) Nature 278: 71
420. Pfohl-Leszkowicz A, Salas C, Fuchs RPP, Dirheimer G (1981) Biochemistry 20: 3020
421. Wilson VL, Jones PA (1983) Cell 32: 239
422. Pfeifer GP, Grunberger D, Drahovsky D (1984) Carcinogenesis
423. Wojciechowski MF, Meehan T (1984) J. Biol. Chem. 259: 9711
424. Ruchirawat M, Becker FF, Lapeyre JN (1984) Biochemistry 23: 5426
425. Boehm TLJ, Drahovsky D (1981) Carcinogenesis 2: 39
426. Boehm TLJ, Drahovsky D (1981) Int. J. Biochem. 13: 1225
427. Boehm TLJ, Grunberger D, Drahovsky D (1983) Cancer Res. 43: 6066
428. Wilson VL, Smith RA, Longoria J, Liotta MA, Harper CM, Harris CC (1987) Proc. Natl. Acad. Sci. USA 84: 3298
429. Drahovsky D, Wacker A (1975) Eur. J. Cancer 11: 517
430. Cox R (1980) Cancer Res. 40, 61
431. Boehm TLJ, Dahovsky D (1981) Cancer Res. 41: 4101
432. Benedict WF, Banerjee A, Gardner A, Jones PA (1977) Cancer Res. 37: 2202
433. Harrison JJ, Anisowicz A, Gadi IK, Raffeld M, Sager R (1983) Proc. Natl. Acad. Sci. USA 80: 6606
434. Hsiao WLW, Gattoni-Celli S, Weinstein IB (1985) Mol. Cell. Biol. 5: 1800
435. Bouck N, Kokkinakis D, Ostrowsky J (1984) Mol. Cell. Biol. 4: 1231
436. Walker C, Nettesheim P (1986) Cancer Res. 46: 6433
437. Yasutake C, Kuratomi Y, Ono M, Masumi S, Kuwano M (1987) Cancer Res. 47: 4894
438. Cavaliere A, Bufalari A, Vitali R (1987) Cancer Lett. 37: 51
439. Carr BI, Reilly JG, Smith SS, Winberg C, Riggs AD (1984) Carcinogenesis 5: 1583
440. Denda A, Rao PM, Rajalakshmi S, Sarma DSR (1985) Carcinogenesis 6: 145
441. Frost P, Liteplo RG, Donaghue TP (1984) J. Exp. Med. 159: 1491
442. Altevogt P (1986) Cancer Res. 46: 2912
443. Walker C, Matthews AM, Shay JW (1987) J. Natl. Cancer Inst. 78: 695
444. Schmahl WP, Toerock P, Kriegel H (1984) Arch. Toxicol. 55: 143
445. Zimmermann FK, Scheel I (1984) Mutat. Res. 139: 21
446. Katz AJ (1985) Mutat. Res. 143: 195
447. Marquardt H, Marquardt H (1977) Cancer 40: 1930
448. von Hoff DD, Slavik M, Muggia FM (1976) Ann. Intern. Med. 85: 237
449. Šorm F, Vesely J (1968) Neoplasma 15: 339
450. Vesely J, Cihak A, Šorm F (1968) Cancer Res. 28: 1995
451. Momparler RL, Gonzales FA (1978) Cancer Res. 38: 2673
452. Vesely J, Cihak A (1977) Cancer Res. 37: 3684
453. Michalowsky L, Jones PA (1987) Mol. Cell. Biol. 7: 3076
454. Reitz MS, Mann DL, Eiden M, Trainor CD, Clarke MF (1984) Mol. Cell. Biol. 4: 890
455. Samid D, Flessate DM, Friedman RM (1987) Mol. Cell. Biol. 7: 2196
456. Lapeyre JN, Becker FF (1979) Biochem. Biophys. Res. Commun. 87: 698
457. Kröger H, Grahn H, Wohlert H (1983) Int. J. Biochem. 15: 627
458. Diala ES, Chea MSC, Rowitch D, Hoffman RM (1983) J. Natl. Cancer Inst. 71: 755
459. Flatau E, Bogenmann E, Jones PA (1983) Cancer Res. 43: 4901
460. Bedford MT, van Helden PD (1987) Cancer Res. 47: 5274
461. Pfeifer GP, Steigerwald S, Boehm TLJ, Drahovsky D (1988) Cancer Lett. 39: 185
462. Feinberg AP, Vogelstein B (1983) Nature 301: 89
463. Goelz SE, Vogelstein B, Hamilton SR, Feinberg AP (1985) Science 228: 187
464. Kuo MT, Bhanumathi I, Wu JR, Lapeyre JN, Becker FF (1984) Cancer Res. 44: 1642
465. Mays-Hoopes LL, Brown A, Huang RCC (1983) Mol. Cell. Biol. 3: 1371
466. Marcaud L, Reynaud CA, Therwath A, Scherrer K (1981) Nucl. Acids Res. 9: 1841

467. Parker MI, Judge K, Gevers W (1982) Nucl. Acids Res. 10: 5879
468. Groffen J, Heisterkamp N, Blennerhassett G, Stephenson JR (1983) Virology 126: 213
469. Catala F (1986) Nucl. Acids Res. 30: 2481
470. Roguel N, Moskowith H, Relevy H, Hamburger J, Kotler M (1987) Biochim. Biophys. Acta 910: 116
471. Kuo MT, Sen S, Tecter L, Cailleau RM, Knesek JE, Bowen J (1984) Cancer Res. 44: 3518
472. Feinberg AP, Vogelstein B (1983) Biochim. Biophys. Res. Commun. 198: 47
473. Ramsden M, Cole G, Smith J, Balmain A (1985) EMBO J. 4: 1449
474. Borello MG, Pierotti MA, Bongarzone I, Donghi R, Modellini P, Della Porta G (1987) Cancer Res. 47: 75
475. Chea MSC, Wallace CD, Hoffman R (1984) J. Natl. Cancer Inst. 73: 1057
476. Kerbel RS, Frost P, Liteplo R, Carlow DA, Elliott BE (1984) J. Cell. Physiol. Suppl. 3: 87
477. Olsson L, Behnke O, Sorensen HR (1985) Int. J. Cancer 35: 189
478. Ormerod EJ, Everett CA, Hart IR (1986) Cancer Res. 46: 884
479. Boon T (1983) Adv. Cancer Res. 39: 121
480. Boon T, Kellermann O (1977) Proc. Natl. Acad. Sci. USA 74: 272
481. Wölfel T, Van Pel A, De Plaen E, Lurquin C, Maryanski JL, Boon T (1987) Immunogenetics 26: 178
482. De Plaen E, Lurquin C, Van Pel A, Mariamé B, Szikora J-P, Wölfel T, Sibille C, Chomez R, Boon T (1988) Proc. Natl. Acad. Sci. USA 85: 2274
483. Scarano E (1969) Ann. Embryol. Morphogen. 1 (Suppl.): 51
484. Barker D, Schafer M, White R (1984) Cell 36: 131
485. Wilson VL, Jones PA (1983) Science 220: 1055
486. Shmookler Reis RJ, Goldstein S (1982) Proc. Natl. Acad. Sci. USA 79: 3949
487. Romanov CA, Vanyushin BF (1981) Biochim. Biophys. Acta 653: 204
488. Wilson VL, Smith RA, Ma S, Cutler RG (1987) J. Biol. Chem. 262: 9948
489. Ono T, Tawa R, Shinya K, Hirose S, Okada S (1986) Biochem. Biophys. Res. Commun. 139: 1299
490. Meijlink FCPW, Philipsen JNJ, Gruber M, Geert AB (1983) Nucl. Acids Res. 11: 1361
491. Holliday R (1986) Exp. Cell Res. 166: 543
492. Fairweather DS, Fox M, Margison GP (1987) Exp. Cell Res. 168: 153
493. Hare JT, Taylor JH (1985) Proc. Natl. Acad. Sci. USA 82: 7350
494. Pages M, Roizes G (1982) Nucl. Acids Res. 10: 565
495. Sturm KS, Taylor JH (1981) Nucl. Acids Res. 9: 4537
496. Adams RLP, Burdon RH, Fulton J (1983) Biochem. Biophys. Res. Commun. 113: 695
497. Feinstein SI, Racaniello VR, Ehrlich M, Gehrke CW, Miller DA, Miller OJ (1985) Nucl. Acids Res. 13: 3969
498. Sanford J, Forrester L, Chapman V, Chandley A, Hastie N (1984) Nucl. Acids Res. 12: 2823
499. Ponzetto-Zimmermann C, Wolgemuth DJ (1984) Nucl. Acids Res. 12: 2807
500. Russell DW, Zinder ND (1987) Cell 50: 1071
501. Abeles AL, Austin SJ (1987) EMBO J. 6: 3185

Author Index Volumes 1–9

The volume numbers are printed in italics